THE
COVENANT
OF THE
WILD

T H E

COVENANT

O F   T H E

WILD

WHY ANIMALS

CHOSE

DOMESTICATION

Stephen Budiansky

ISBN 0-9648750-0-4
Previously published by William Morrow and Company
 (ISBN 0-688-09610-7)

Printed in the United States of America

First Terrapin Press edition 1995

1 2 3 4 5 6 7 8 9 10

Cover design by Martha Polkey

PREFACE

In the summer of 1981, possessed by visions of serene sunsets and green fields of clover, and buttressed by a dozen pounds of U.S. Department of Agriculture publications, I rented a small farm in rural Maryland. The clover—after two years, a drought, a torrential rainstorm, a broken tractor axle, and regular weekend visits to farm-supply dealers—finally did turn green. (The dealers, though, turned out to be masters at playing on the insecurities of a suburban kid trying to pass himself off as a real farmer, especially a suburban kid who grew up petless and gardenless outside of Boston. I took to wearing one of those farmer's baseball caps with a feed-company logo on the front wherever I went, a pathetic disguise that fooled no one.)

But the sunsets faded in the first few months and, a decade later, still have not fully emerged from behind the clouds of tragedy that, as any real farmer knows, are always hovering over a barnyard. Hens were massacred, necks broken but otherwise left untouched by a marauding weasel that slipped through a one-inch hole in a carefully fenced coop; hatching goslings died exhausted in their pipped shells; and sheep, traitors all to the pastoral muse, bred worms and lambing-time disasters in equal measure. Friends would say how wonderful it must be to live on a farm. I would reply it was, but there were lots of head- aches, too. Never believing that I really meant it, they would only compliment me further on my rustication, mar- veling that I had even learned to talk like a farmer— forever complaining in the time-honored fashion.

No matter how I steeled myself, no matter how many gory descriptions of disease and obstetrical complications I had studied in morbid anticipation ("sometimes there are just clots, often blood or a thick yellow puslike substance that is difficult to milk out"), and no matter how firmly I placed my feed-company cap on my head as I trudged out to the barn, the truth was that the suffering of animals I was forced to see with such alarming regularity on a farm rent my heart.

When I had just about decided that I would never make it as a real farmer, I discovered something amazing: These things rent the hearts of real farmers, too. Somehow, to face every day the brutality of nature wasn't necessarily brutalizing. Sending animals off to the butcher didn't cover one with a layer of emotional calluses to match the ones on one's hands. And the reasons turned out to be simple, and, in hindsight, obvious. Farmers dealt with tragedy not by closing their hearts to the suffering within nature but by opening their minds to the larger truth of nature. Farmers understood, even in their little tame corner of the natural world, that nature is a force larger than themselves, with its own rhythms, its own purpose, its own sense of morality that makes a mockery of man's. They still cared and strug- gled to save the dying lamb, in spite of the seductive ni- hilism that whispers why bother, it is born to die soon anyway. And when the time came for it to die, they still packed it off to market, in spite of the seductive sentimentality that would make those of us less wise, but no more humane, do otherwise. It was a lesson in how much we have simply forgotten in our civilized lives, where the harsh rules of nature almost never intrude. We have forgotten how to face life without losing a grip either on our humanity or on reality.

This book was born in part of my experiences as an impostor farmer; in part of my fascination with archaeological and behavioral research that is beginning to reveal domestication of animals as a natural product of evolution; and in part of my anger at the simplistic stereotypes of man and nature that are being purveyed by an ever more confrontational animal rights movement to an ever more urban audience. In a day when many of the animal rightists' claims about the supposed cruelties of farming have attained the status of conventional wisdom (everyone seems to know about the suffering of veal calves and laying hens), few of us have the true wisdom or experience to place these claims in context. Ignorant of the realities of nature, we imagine a paradise where all is well but for the cruelties inflicted by man. It makes for simple choices and easy moralizing, something that is rarely compatible with experience. As essayist and wounded World War II veteran Paul Fussell said of a critic who denounced the atomic bombing of Japan, "I don't demand that he experience having his ass shot off. I merely note that he didn't." There is a similarly galling naïveté in those, far removed from nature, who see in farming the "exploitation" of animals and who imagine that anyone who disagrees with them is either the unenlightened slave of unquestioned convention or an economic creature of base self-interest. I don't demand that those who would dictate to farmers how to treat

their animals lie on a cold barn floor with their hand up a
ewe's uterus to save the lives of ewe and lamb; I don't
demand that they see a goose beheaded by an owl; I merely
note that they haven't.

Understanding need not come only from experience. A
wider appreciation for the role played by the natural forces
of evolution in the ancestry of domesticated animals may
help to add much-needed depth to the often one-
dimensional discussions that carry the day. If, rather than
just another instance of man's arrogant exploitation of nature, domestication is instead a product of nature, then we
will have to think more carefully about the interconnections between all species, and be less quick to apply the glib
slogans of human politics, the language of "rights" and
"exploitation" and "oppression," to relationships crafted by
forces in many ways beyond our control. We cannot all be
farmers, but we can all be students of evolution. As more
of the natural world comes under our control, we had
better start understanding how the natural world really
works, one way or another.

The ewes in my barn have begun to lamb this year
again, so far with only minor tragedies. I have even become
emboldened, and look in on them feed-capless. Slowly,
even some of the pastoral delusions that first beguiled me
to the land are finding a place beside the harsher realities.
There's a simple pleasure and beauty in seeing a minutes-
old lamb stand up on his wobbly legs to nurse; it's a pleasure made deeper by knowing that while it cannot last
long, it will come again next year.

—S.B.
Mountville, Maryland

ACKNOWLEDGMENTS

All journalists are parasites, but science writers are a special case. They are what biologists would term social parasites—organisms that make a living mimicking the behavior of another species. Science writers know just enough about the language and conventions of the scientific world to elicit the responses a researcher instinctively gives to a colleague or a graduate student—that is to say, to a real scientist. The writer is given patient explanations, reprints of journal articles, helpful suggestions for lines of further inquiry. It is only afterward that the scientist discovers the writer's real motive is not the advancement of science, or even the advancement of knowledge, but merely the advancement his own necessarily selfish ends.

As a social parasite with a social conscience, I thus feel that the very least I can do is to acknowledge the many scientists who have shared their time and thoughts with me, and without whose unsuspecting generosity this book could not have been written. I was especially helped by Raymond Coppinger of Hampshire College in Amherst, Massachusetts, whose studies on the domestication of dogs—a remarkable blend of the practical and the theoretical—are the basis of a great deal of what appears in these pages. His work in breeding livestock-guarding dogs and his extraordinary paper "The Domestication of Evolution" first acquainted me with the idea that domestication could be considered a natural evolutionary process, and first started me thinking about the implications of this idea.

David Rindos of the University of Western Australia and Andrew Rowan of Tufts University freely shared with me many invaluable insights; Melinda Zeder of the Smithsonian Institution set me in the right direction by detailing for me an entire history of the study of agricultural origins and by providing numerous leads.

I would also like to thank the many other scientists I have spoken with over the years whose ideas have found their way into these pages: Darcy Morey of the University of Tennessee; Michael Robinson of the National Zoological Park in Washington, D.C.; David Steadman of the New York State Museum; Julio Betancourt of the U.S. Geological Survey; Anthony Legge of the University of London; Dr. Gerald Hart of the University of Toronto; George Armelagos of the University of Massachusetts; Donald Ortner of the Smithsonian Institution; and Juliet Clutton-Brock of the British Museum.

In the strange world of news magazine journalism, "serious" science stories (those that fail either to predict the immediate end of the world or to mention cholesterol at least once) are a hard sell; I am grateful to Roger Rosenblatt, Mike Ruby, and Mimi McLoughlin at *U.S. News & World Report* for encouraging me to write the story that this book grew out of and for so flouting the conventional wisdom as to risk putting it on the cover.

Finally, I should be remiss if I did not thank Mary Taylor, who taught me to think like a horse, and my wife, Martha, who really started it all by coming home one day with a pair of Pilgrim geese.

CONTENTS

THE
COVENANT
OF THE
WILD

I

VISIONS
OF NATURE

Civilization has so cluttered this elemental
man-earth relation with gadgets and
middlemen that awareness of it is growing
dim. We fancy that industry supports us,
forgetting what supports industry.
—Aldo Leopold, A Sand County Almanac

For two million years we were hunters; for ten thousand years we were farmers; for the last one hundred years we have been trying to deny it all, except for the fervent moments when we try to recapture it all, or do both at once. To an ever more urban America, animals are things, antiseptic packages of featherless and hairless chicken wings in a plastic wrapping on a Styrofoam tray; or they are people, a beer-drinking dog in a television commercial. The true character of animals and their meaning in the world, once common knowledge to the humans whose lives intertwined with theirs, is today lost in a miasma of human fantasies.

If the Industrial Revolution made animals into mere objects to be used as man saw fit, the nature-worshiping counterrevolution that followed made them into objects of adoration to be revered. Wolves, the favorite villains of fairy tales, are now ecological heroes; but in transforming them into majestic symbols of the wild and free, today's nature enthusiast is no closer to understanding their true nature than were the brothers Grimm. The favorite animal in a survey of visitors to the National Zoo in Washington, D.C., was the giant panda, typically described by zoogoers as "cute, cuddly, and adorable." It actually is solitary, ill-tempered, and aggressive, but never mind. Some zoologists suggest that the influence of generations of teddy bears, combined with the panda's stubby limbs, big head, and seemingly big eyes, appeal to our naive sense—innate or learned—of parental protectiveness. Mere facts cannot compete with perception.

Genuine understanding, which used to come from actual experience of the natural world, and which today could come from scientific studies of behavior and ecology, is no match for such human preconceptions in a world where nature is a factory, a shrine, or a theme park. The timber-company manager sees a stand of old-growth forest as a product, and the spotted owl as a manufacturing problem; the Sierra Club loyalist sees the forest as a sanctuary defiled by man, and the spotted owl as a sacred icon; the tourist—well, he'd just as soon go to Yellowstone and have his picture taken standing next to a bear.

If we grasp so little about the true nature of nature, a world where wolves tear the hindquarters off living sheep, where pandas snarl, and where elk starve to death, it is hardly surprising. We still depend upon animals as much as did those distant ancestors of ours who, crouching in a torch-lit cave in what is now Lascaux, France, painted portraits of the horses, cattle, and deer that lived alongside them on the barren tundra of Ice Age Europe. We depend upon them as did our not so distant ancestors who beseeched agricultural relief officials during the Great Depression for feed for their starving cattle and horses. Yet the knowledge is gone. In 1880, one in two Americans lived on a farm; today, one in fifty does. Even in 1930, within the living memory of many Americans, one in four lived on a farm, and most others were but one step removed. Many had grown up on farms. The plight of farmers was national news. Companies that supplied farmers found it worthwhile to advertise on radio networks. Even for many nonfarmers, chickens came from the backyard. Even when they came from the market, they were undisguised by plastic and Styrofoam. They were chickens, with feathers on the outside and intestines on the inside.

The urge to turn animals either into things or into people reflects the distance we have traveled in a generation or two. We conveniently alternate between anthropomorphism and blindness. Gourmet dog food is a great marketing success, since people, not their dogs, do the shopping; what we in the United States spend on our dogs and cats, observes zoologist Michael Robinson, amounts to a sum greater than the entire economy of medieval Europe. Meanwhile, ten million abandoned dogs and cats are put to death every year in animal shelters across the country. And, meanwhile, impassioned protests are mounted against the use of two hundred thousand dogs and cats in biomedical research every year. The terrible difficulties in balancing on the one hand the real ability of advances in biomedical research to alleviate pain and suffering, and on the other the real pain and suffering that is inflicted upon at least some research animals in the process, are all too easily swept aside in favor of stereotypes: Research animals are either hapless victims, to be rescued in daring commando raids and bestowed with names, or they are pieces of laboratory equipment, ordered from a catalog along with the Petri dishes and pipettes. It is simply too hard, it seems, to see them for what they are—animals that can suffer, but animals that are not people, either.

The challenge for society carried ever further from personal acquaintance with nature is to retain a grasp on its place *in* nature. Yet the tendency of both those who would conquer nature and those who would save it is to draw a line and declare human society ever *apart* from nature. The Victorians, starry-eyed over steam engines and railroads, saw in nature just another tool. "If there is anything in the world of nature that seems clear, morally," wrote one Victorian thinker, "it is that man has an authentic right to require reasonable service from the horse." Today's environmentalist, cross-eyed over sulfur-dioxide-belching factories and oozing drums of noxious chemical waste, sees in nature an innocent victim of human greed. And the majestic creatures of the wild—wolf, whale, seal, bear—are the emotionally powerful symbols of this violated Eden.

The environmental movement's view of nature has become the predominant one in America today—nature as a pristine world, spoiled only by modern man, the klutz in heavy boots trampling the flowers. To find any reality in the vision of a nature separate from man, though, we would have to turn back the clock at least ten thousand years, back before the Paleo-Indians of North America began running herds of mammoths off cliffs to their extinction, before the first slashing and burning agriculturists leveled the forests of Europe and Asia, and before the evolutionary imperative of domestication spread dogs and cows and sheep, and man, across the face of the earth.

The truth, as ecological research struggles so often in vain to tell us, does not admit of any simple lines between that which is man's and that which is wild. The interests of people and other animals do collide; that which is wild and free is not always heroic and beautiful. Man has always been a part of nature, a fact we can ignore now only at the peril of man, animals, and the world.

For how we think about these things matters as it never has before. Even as we as individuals have lost touch with the realities of nature, we as a species have become ever more a force in nature. We are partners in a new evolutionary dynamic that now dominates the earth— domestication. We are also stewards of a land altered forever by that dynamic, as our fields and houses overrun the habitat of countless species, from the panthers of Florida to the rhinoceroses of Africa.

Yet faced with the challenges of preserving wildlife habitat, protecting the earth's biodiversity, and saving endangered species, we devise solutions based on a model of nature that is ten thousand years out of date and somehow expect them to work. We believe that all we need to do is leave nature alone, and do no overt harm, and everything will be fine, as if our very existence had not irrevocably altered the ecological balance of the planet.

With so much at stake, an artificial conception of nature is not just naive: It can be fatal. At Yellowstone National Park, protected herds of elk and bison that grew to record numbers during a series of mild winters and wet summers severely overgrazed the range. But when the inevitable crash in populations occurred in the harsh winter of 1988–89 that followed the spectacular 1988 Yellowstone fire, televised images of hungry elk brought hundreds of angry letters to park officials demanding that the animals be fed. Never mind that human intervention would only further destroy the overstressed habitat, increase erosion, and drive out moose, bighorn sheep, grizzly bears, and other competing fauna; never mind that winter feeding would artificially extend the natural life span of the elk and bison populations past the point where the ecosystem could sustain them; and never mind that the six thousand elk and four hundred bison that did starve to death in the park provided a bounty for the scavengers, such as eagles, coyotes, and bear, that are also a part of a healthy ecosystem. The conception of nature that sees everything bad as man's doing and thus man's responsibility is a powerful force. Yellowstone's superintendent, Robert Barbee, and others have complained of the attention lavished upon "charismatic megafauna" at the expense of the ecosystem that sustains all life in the park, and are doing what they can to educate the public. ("Is This a Tragedy?" asks one poster displayed at the Old Faithful visitors' center; it shows the carcass of a bison, and goes on to explain the role of scavengers in the Yellowstone ecosystem.) But, still, outraged visitors recently demanded to know why park officials did nothing to help a hapless bison that fell through the ice in full view of tourists.

Many biologists argue that in the absence of the predators—both Indian hunters and wolves—who had once kept the elk populations in check, overgrazing will continue; they have suggested that park rangers shoot animals that exceed the natural carrying capacity of the land.

Given the outraged protests that greeted the shooting of 569 bison that strayed over the park's boundary into Montana in the winter of 1989–90 (the hunt actually led to an emergency congressional hearing), any such plan to shoot animals within the park seems beyond political possibility. Now that overgrazing elk have so severely reduced the aspen that beavers need for food and shelter, and beaver numbers have dropped, the environmentalists' own favored option, of returning wolves—who need small game such as beaver as well as large game such as elk to sustain them— may be foreclosed, too.

Those who would save the endangered African elephant, charismatic megafauna if there ever was one, may likewise be killing it with naive kindness. A worldwide ban on the ivory trade adopted in 1989 may have exactly the opposite effect from that intended, largely because of a failure to consider the possibility that African villagers might be part of a functioning ecosystem. The elephant-savers' appeals focused on brutal poaching by heavily armed gangs in Kenya's national game parks, where elephant populations dropped from 65,000 in 1979 to 19,000 just ten years later. All but ignored, however, was Zimbabwe's success in enlisting farmers and villagers in a program of controlled hunting, which has resulted in an *increase* in that nation's elephant population from 30,000 in 1960 to 52,000 in 1989. Villagers shared in the proceeds from the sale of ivory from culled elephants and the $15 million a year brought in from wealthy sport hunters who were willing to pay up to $30,000 apiece to shoot one of the animals. In a part of the world where human demands for land are increasingly in conflict with the needs of animals, the program actually resulted in farmers' setting aside land for elephant range because they found it more profitable than growing crops.

But the trade ban adopted at the urging of animal-welfare organizations, which includes a ban on importing whole tusks as trophies, has threatened to destroy the successful Zimbabwe effort, while perhaps only driving up the price of black-market ivory and thus making poaching even more profitable.

The sentiment that prompted the European Economic Community in 1983 to ban imports of seal pelts equally overlooked the fact that the Inuit of Canada, who had made their living for centuries fishing, hunting, and trapping, were no less a part of the ecosystem than were the seals. Now suffering the consequences of the larger antifur movement sweeping the United States and Europe, the Inuit tell of a drastic increase in suicide, alcoholism, and drug abuse as their traditional way of life vanishes. The Reverend John Sperry, Anglican bishop of the Diocese of the Arctic, told the *Christian Science Monitor* that the Inuit "can't understand why people in southern climes who kill millions of animals for their tables and for their shoes are attacking the only economic means they have to stay off welfare."

The confusion and inconsistency in our modern, urban vision of nature in these "southern climes" has proved fertile ground for at least some of the ideas of the animal rights movement to take root. It is a movement that is easy to underestimate. Its basic philosophy certainly has little broad appeal. Peter Singer, the Australian philosopher whose 1975 book *Animal Liberation* has become the bible of the movement, sets forth an absolutist position that most people simply find morally incomprehensible. To Singer, all animals that are "sentient"—aware and able to experience pain—deserve equal consideration. To place a greater value on the interests of one's own species at the expense of another's is "speciesism," analogous to racism. Singer thus argues that if we could not justify performing a scientific experiment on a severely retarded human, we could not justify performing it on a nonhuman animal. Raising animals in captivity or keeping them in laboratory cages is slavery, morally unacceptable no matter what the purpose. A respect for the rights of nonhuman animals requires an end to all animal experimentation, hunting, and animal agriculture.

The many high-visibility protests and civil disobedience actions by animal rights groups have probably not done much to broaden their appeal, either. When the Humane Society of the United States, a major animal protection organization, issued a press release in 1989 explaining why it had joined the fight to save the world from the greenhouse effect ("Animals will be among the first to suffer in a deteriorating environment. Unable to reach greener pastures and cool, clear waters, animals will roast inside their fur coats, broil in the burrows and suffocate in lakes and oceans"), it seemed at best an exercise in self-parody. Or when the animal rights group People for the Ethical Treatment of Animals bought forty dollars' worth of lobsters from the tank of a Chinese seafood restaurant in Rockville, Maryland, and then spent two hundred dollars to transport them on a United Airlines flight to Portland, Maine, so that they could be returned to their native waters, it was difficult to view them as much of a threat to anyone or anything.

Even break-ins, bombings, and death threats by the shadowy Animal Liberation Front and other so-called direct-action groups have probably done more to discredit the movement than to achieve their stated aim of halting animal experimentation. The break-ins have, to be sure, succeeded in driving up the costs of research: A fire set by the ALF destroyed an animal diagnostics laboratory under construction at the University of California, Davis, at a cost of $3.5 million, and a series of attempted break-ins at the Yerkes Regional Primate Research Center in Atlanta, Georgia, have forced administrators to divert several hundred thousand dollars from research to security. But such acts of vandalism have also galvanized the scientific community to become much more outspoken in defending their work and in bringing that message before the public.

The net effect of the animal liberators, arguably, has been only to put off the day when researchers and regulators themselves begin to weigh the true ethical cost of animal experimentation and testing, and take upon themselves the moral responsibility of questioning whether the need is so great that it justifies that cost. The extremism of the abolitionists has tarnished the moderate reformers—those who seek improved care and handling of laboratory animals, the development of alternatives (especially in the case of routine safety testing of consumer products, where alternatives show the greatest promise), and the use of the minimum number of animals in areas of basic research where there are no good alternatives. It has become all too easy for researchers to dismiss any criticism as Luddite and misanthropic. The more strident the movement as a whole becomes, the less effective it is at achieving its direct aims.

Yet for all its arid philosophizing, public buffoonery, and underground terrorism—a combination that does not exactly add up to a winning public relations strategy—the animal rights movement has been extremely effective in an indirect way, insinuating its simplistic views toward animals and the natural world into the public's consciousness.

Some of these views are sincerely held, but some appear to be a cynical attempt to advance an unpopular, abolitionist agenda via emotional appeals that the animal rightists know will strike a responsive chord in a public that knows little of the hard choices nature poses. When antihunting activists say they are trying to save the deer (or better yet, the fawns), few of us ask whether the deer would prefer starvation. When antifarming activists say that confining cattle to the farm dooms them to a life of boredom, few of us ask whether the cattle would have preferred extinction. When Singer argues, in *Animal Liberation,* that all we need to do to protect the interests of animals is to stop such aberrant behavior as hunting and experimenting on animals because "in our normal life, there is no serious clash of interests between human and nonhuman animals," few of us calculate the much greater toll that our mere existence—our roads, houses, cities—has taken on wildlife habitat. It all makes for simple choices: The way to save animals is to be nice to them.

The ready equation of "natural" with a romantic, twentieth-century urbanite's vision of nature, one in which elk never starve and cows graze happily in pastures green, has already had very real policy consequences in Europe. Sweden passed a law in 1988 actually requiring cows, pigs, and animals raised for fur to be kept "in as natural an environment as possible," begging the question of what is natural for animals that are by their evolutionary heritage incapable of surviving in the wild. It is unnatural to feed a cow hay in the winter, for example; it is unnatural for cows even to *be* in Sweden in the winter in the twentieth century, for that matter. Wild cows became extinct in Europe thousands of years ago. Still, the law specifically requires that cows must be allowed to graze on pasture in the summer. In a bow to reality, it doesn't mind if cows are shut up in a warm barn in the winter.

Similar laws have been proposed in the United States, and although none has yet passed, more than one hundred thousand Massachusetts residents signed a referendum petitions to place such a proposal on the state ballot in 1988. It failed, much to the relief of state agricultural officials who predicted with little exaggeration that the law's effect would have been to end farming in the state.

Playing on the same sentiment evident in the controversies at Yellowstone, however, animal rights groups have already made considerable legal inroads against hunting. Voters in California approved a referendum measure in 1990 banning mountain lion hunting; legal maneuvering by activists in Michigan, Ohio, and New York have succeeded in blocking the introduction of dove hunting in those states.

If all of these moves are incomprehensible to hunters, farmers, and the few others in our modern world whose daily work still brings them into contact with animals, it is not because they are without compassion. It is because they know better. They know about the interdependence and competition that tie the species of this planet together, because they see it every day. They know where meat comes from; they know, too well, about diseases, and sheep-killing coyotes, and survival. They know that death is a part of life, that in nature death makes life possible. They know that domesticated animals need us as we need them. They daily experience the wonder of our shared evolutionary heritage with animals—the dogs that have so long served man as hunter, herder, and guard; the cattle and sheep, in whose dependent tameness the foundations of agriculture, and in turn a civilization of laws and learning and science, still ultimately rests. They have seen, simply, something larger than ourselves that we, whether we know it or not as we go to work in office buildings and eat food from plastic wrappings, are still a part of. They know that animals are not people, but they are not things, either. They know, in short, the contract that was sealed when man and domesticated animals cast their lot together ten thousand years ago.

Scientific insight has in the last century reaffirmed and expanded much that farmers (and hunters before them) have known all along. Farmers and hunters were, after all, in a very real sense the first animal behaviorists. The cave paintings of Lascaux were the work of people who knew the form and movements and habits of the animals upon whom their livelihood depended.

Yet as such personal experience became less and less a part of a larger public's consciousness, formal knowledge in the form of scientific observation and record maddeningly failed to take its place. Even the revolutionary ideas of Charles Darwin, whose 1859 opus *On the Origin of Species* was what today would be called a best-seller, were interpreted more through the lens of his readers' preconceptions than for what they plainly said. They were taken by the antireligious as disproof of the existence of God, by the religious as a threat to man's belief in God, by imperialists as proof of the white man's burden, by conservatives as justification for the social order.

Darwin attempted to prove none of these things. He said only that nature is not a perfect, harmonious stasis but rather a turbulent, dynamic engine of adaptation. He explained why animals do what they do—not because they are good or evil, heroes or villains, noble or base—but because they are adapted to their environment. And he planted a seed for understanding the origin of the complex relationship between man and domesticated animals, a seed that has flourished in the last two decades with behavioral and archaeological evidence that points to the cooperative evolution of our species in a mutual strategy for survival.

If we can understand that evolutionary heritage, we can understand much more. If we come to see the domestication of animals not as a crime against nature, but as a product of nature, we will in the process confront something much more fundamental about man's part in nature. It is impossible to reconcile the environmentalists' view of a nature forever separate from man with the evolutionary facts of domestication. If life with man was a better evolutionary bargain for domesticated animals than was life in the wild, then it makes no sense to say that nature (really just another word for evolution) ends where man's presence begins. And it raises doubts about larger judgments based on the premise that whatever is wild is pristine, whatever is human is tarnished. We are easily shocked by the horror stories of the laboratory and barn in part because we are ignorant of the greater horrors of the wood and water, horror stories written by nature herself.

The scientific story of domestication is an allegory about nature in another sense—it is a blunt reminder that nature, that is to say evolution, does not really care about that thing that so inflames visitors to Yellowstone, namely the plight of the individual bison falling through the ice. All of nature's strategies for survival of a species, strategies that include domestication, include suffering and death of individual members of that species. Old moose fall to pursuing wolves in the wood so that the young might live; lambs die to an axfall behind the barn so that more don't die of starvation, hunger, disease, and predation. Now that we have, willingly or not, become stewards of so much of nature, it is a lesson we had better learn. To save a few elk from starvation in a harsh winter is to do them no favor.

The story of domestication that follows is, like the best scientific stories, counterintuitive. Indeed, most of us are so accustomed to the notion that domestication was a human exploit that to suggest otherwise can make one sound more like a mystic than a scientist. There are some fundamental scientific concepts, however, that underscore the argument for domestication as the evolutionary product of a mutual strategy for survival, and setting these out at the start may help the reader approach the subject with an appropriate measure of doubt about the conventional view of things.

First, the ancestors of today's domesticated plants and animals were, like mice and starlings invading our houses today, opportunists, not conservatives. They were species adapted by their evolutionary past to exploiting new terrain on the forest edge, rather than specializing in niches in the forest center. The first domesticated animals—dogs, sheep, and cattle—were social species that readily scavenged new food sources. The first domesticated plants found in North America were, similarly, weeds, in effect—sunflowers and gourds that readily invaded any disturbed ground.

The story told again and again in the archaeological record is one of long, loose associations between free-living partners before a full-fledged domesticated relationship appears. The rise of agriculture, long viewed as a revolution, was really a long evolution. Man and animals and plants engaged in thousands of years of close association before domestication blossomed. At Tepe Ali Kosh, an early food-producing site in what is today southwestern Iran that has yielded some of the best archaeological data, domestic plant species account for 5 percent of the seeds from nine thousand years ago; a thousand years later they have grown only to 40 percent. At Tel Abu Hureyra, one of the first permanent settlements in the world, located on the banks of the Euphrates River in northern Syria, bones of domesticated sheep and goats appear ninety-five hundred years ago; they account for 10 percent of the bones found at the site (the rest are almost entirely wild gazelles) until finally growing to 60 percent around eighty-five hundred years ago. Wild cats, which probably invaded the granaries of the first agricultural settlements of the Near East some nine thousand years ago in pursuit of rodents, lived in free association with humans for thousands of years before the Egyptians, who associated the male cat with the sun-god Ra and the female with the fertility goddess Bast, began confining them to temples and deliberately breeding them, around four thousand years ago. Over the next several thousand years, they then spread to other parts of the Near East and then into Europe.

Another consideration that makes the story of domestication less surprising and more understandable is the infinite genius of evolution in devising survival strategies that involve cooperation among species throughout nature. Man is far from the only species to practice domestication. Cooperative associations between some unlikely pairs— finches and wasps, ants and trees, aardvarks and melons— appear throughout nature. In almost all, there is one recurring pattern: The defense mechanisms that allow a species to survive on its own, but likewise make it fearful of associating with others, are dropped; in return, tangible benefits in the form of protection or food are gained. The state of dependence of one species upon another so formed is not degeneracy; it is a finely honed evolutionary strategy for survival. In a world made up so much of competition for survival, nature has with surprising frequency cast upon the solution of cooperation.

Finally, all domestic animals, in both behavior and appearance, retain juvenile traits in adulthood. One of the very first hints in the archaeological record of an animal's domestication is the jawbone of a wolf from southwest Asia, dated twelve thousand years ago, in which the face and muzzle have begun to shorten—an adult with the face of a puppy—crowding the teeth together. It is a process that has been repeated in every domestic animal. And that one fact of their evolution speaks volumes about what they are and how they came to be. Domestic animals are dependent, permanently juvenile, by nature, not just by circumstance or training. And, according to fossil evidence that long predates the age of domestication, they are completing a process of genetic evolution that was initiated not by man's captive breeding, but by the need to adapt to an environment racked by waxing and waning glaciers in the Ice Age that lasted a million years and ended only ten thousand years ago.

An understanding of these scientific discoveries may be an antidote to the delusions of the last century—a century that has seen the keepers of our ancient contract with domesticated animals dwindle and their ten-thousand-year-old legacy mocked, distorted, manipulated, and forgotten. It is an antidote to an image of a nature that never was, an Eden whose impossible standards man will never live up to, but in trying to will neglect his more worldly obligations.

II

CIVILIZATION'S
PROGRESS;
OR, WHO INVENTED
THE DOG?

*I am wont to think that men are not so
much the keepers of herds as herds are the
keepers of men, the former are so much the freer.*
—HENRY DAVID THOREAU, Walden

I t is a central myth of our culture that we are firmly
in control of our destiny. So the first chore for anyone who would set out to prove that the domestication of
animals is a natural product of evolution is to undo several
thousand years of human self-importance.

It is no easy task—one compounded by our ignorance of
the animal world and the science of evolution. The Citizens Humane Commission of Berkeley, California, a city
that can always be counted upon to encapsulate America's

loss of knowledge about the real world, not long ago conducted a spirited debate over exempting miniature pet pigs from an ordinance banning farm animals within the city limits. The issue, a spokesman told the *Wall Street Journal,* was not concern over whether the pigs posed a threat to hygiene, or whether they would create a nuisance to neighbors, but "whether it is humane to impose domesticity on yet another animal." One commissioner argued that domestication of any animal "deprives that animal of living its own life," and vowed to fight the commission's mildly pro-pig recommendation.

The limitless power of human invention is something we simply take for granted; it seems obvious that we "imposed" domestication on animals, just as we imposed our will in countless other ways on the world around us. In one of his "Two-Thousand-Year-Old Man"* routines, the comedian Mel Brooks recounts how a caveman named Bernie first "discovered" woman. ("I think there's ladies here," Bernie announces one morning). As a commentary on the often exaggerated regard we hold for human accomplishments in the history of the world, it's not too far off the mark. We see ourselves having overcome the forces of nature—fashioning tools, crossing oceans, eradicating diseases—through individual acts of heroism and ingenuity.

In recent times, especially in America, the myth has been extended, often ruthlessly, to the most trivial aspects of an individual's success or failure. As the historian Robert McElvaine notes, the American belief of individual responsibility added a devastating psychological burden of guilt to

---

*A case of poetic license: To have fraternized with "cavemen," Brooks's character probably should have been the Twenty-Thousand-Year-Old Man.

the physical want of those who lost their jobs in the Great Depression. Having taken credit for their modest prosperity in the 1920s, they were sure that there was something wrong with them when they became "failures" in the 1930s. The undeniable power of human invention in this century to alter or even dictate the terms of history has only reinforced the conviction that, both as individuals and as a species, we are in command.

Even without such tangible evidence as atomic bombs, there is a very natural human tendency to explain the world through the experience of man. Even primitive peoples who live on more intimate terms with forces beyond their control than do most of us today clearly have a long tradition of casting themselves as the central figures, if not the stars, in mythical explanations of the things around them. Their traditional stories of how animals were domesticated are no exception. The seminal event in these tales is a person's going into the bush and bringing back a young animal and rearing it in the village. Typical is a campfire story from southern Nigeria, as related by the zoologist Frederick Zeuner:

> A boy adopted a wild dog's pup, grew fond of it and brought it up in the village in spite of the attempts of the pup's mother to rescue her child. When fully grown, the dog induced a bitch to join him, and their litter became used to camp conditions immediately. They went out on hunting expeditions with their human friends. Subsequently the inhabitants of other villages imitated the practice.

It is only a folktale, of course, and perhaps we shouldn't read too much about human attitudes into tales told around the campfire to pass the night. Except for one striking fact: This folktale is almost indistinguishable from the serious explanations offered in most contemporary studies of domestication. The portrayal of domestication as an idea that struck Bernie one fine Mesolithic morning has had a remarkable resiliency in the scientific literature. Treatises of otherwise impeccable scholarship, filled with pages' worth of painstaking analyses of fossil teeth and ancient human demographics, suddenly revert to a Just-So story when it comes to the crucial question of *how* domestication occurred. A typical example: "It may be surmised"—the passive voice is a warning that we're exiting the realm of science—"that wolf puppies were occasionally taken from dens by women and children who fed and played with them." The unquestioned assumption is that the actual process of domestication is trivial; the only real challenge is to explain why someone was motivated to come up with the idea in the first place.

The idea that domestication had its origin in the practice of adopting pets from the wild was reinforced by studies of modern-day primitive peoples. Until detailed archaeological studies began to prove them wrong, anthropologists long took it for granted that modern primitive peoples were the exact counterparts of ancient primitive peoples. Thus they assumed that by studying the habits of the North American Indians or the Australian aborigines, societies that had not yet developed animal agriculture, they could discover what had happened thousands of years ago in Europe or southwest Asia at the very dawn of domestication.

Early European explorers of North America observed that the Indians kept an astonishing variety of pets. Tame raccoons, bears, wolves, even moose and bison, lived in

Indian encampments. The travelers described the affection the people lavished upon these pets, and how difficult it was to persuade the Indians, especially the women and children, to sell or trade their animals. The women, according to some of these accounts, even suckled the young animals from their breasts.

In this century, South American Indian tribes have been observed to keep tame monkeys, rats, parrots, macaws, and jaguars; Australian aborigines often bring back wallabies, opossums, birds, and frogs from the bush. In the case of the aborigines, though, most of the animals are tied up in camp and quickly die from inadequate care and feeding.

Can we be seeing in these modern vignettes a reenactment of the first crucial steps toward domestication some ten thousand years ago? It is tempting to think so. But there are two fundamental problems with trying to explain domestication as the product of nothing more than humans' having tamed a few animals from the wild.

The first is the extraordinary high failure rate of man the domesticator. Yes, Indians kept moose, raccoons, and bears as pets, but not one exists as a domesticated species today. The ancient Egyptians, whose very civilization was based on cattle herding and who were well versed in the mysteries of animal husbandry, tried but failed to domesticate gazelles, ibex, hyenas, and antelope. Egyptian art depicts them and other animals with collars around their necks or being herded. Yet even for this highly developed agricultural civilization, such experiments led to nought. By contrast, several thousand years earlier, the very first agriculturists, people who had never built a fence or mowed a hayfield, succeeded in domesticating virtually every animal that even today, more than five thousand years later, occupies a place of importance in our homes and fields.

Why should sheep and cows have been domesticated but not gazelles or antelope? If it was not for want of human effort, then the explanation must lie elsewhere—in the realm of biology, not anthropology.

The second problem with equating domestication with taming is that the very traits that would have made a plant or animal even amenable to domestication or profitable from a human standpoint—docility, lack of fear, high reproductive rate—were simply not present in the wild type that the Mesolithic hunter first encountered. If we are to believe that domestication was the result of human exploits alone, we run into biological paradox: The only way to produce an animal with the desirable traits is through captive breeding, yet the only way they could have been captively bred is if they had the desirable traits to start with.

This paradox is the crux of the entire, counterintuitive line of evidence that argues for domestication as an evolutionary, rather than a human, invention. The only way out is to recognize that, in an evolutionary sense, domesticated animals chose us as much as we chose them. And that leads to the broader view of nature that sees humans not as the arrogant despoilers and enslavers of the natural world, but as a part of that natural world, and the custodians of a remarkable evolutionary compact among the species.

Why has this paradox been so long ignored—not just by spinners of folktales but even by science? The short answer is that domestication has largely been studied as an anthropological issue, a matter of cultural change. At its heart, this may be a reflection of the natural human tendency noted above to cast human actions as paramount. But the assumptions made by the first practitioners of the young science of archaeology dramatically reinforced that tendency, establishing a pattern of scholarship that is still strong today. Being only human, the first archaeologists assumed that human action was what mattered. Being only European and nineteenth-century humans, they assumed that progress, and especially the progress that led humans to abandon hunting in favor of animal husbandry, was so natural an aspect of human behavior that it required no explanation.

In the late eighteenth and early nineteenth centuries, the first archaeologists were beginning to unearth stone and metal artifacts of earlier human societies, and in an effort to make sense of a growing body of evidence, they developed a simple and powerful model. It was the Danes who first proposed that stone, copper, and iron artifacts represented a series of cultures of increasing technological sophistication. By the time the Danish National Museum opened its doors in 1819, the idea of three distinct periods of prehistory—a Stone Age, a Copper Age, and an Iron Age—was firmly enough established that it was used as the ordering principle for the antiquities collections. It quickly was adopted in other European countries as well.

As an intellectual tool, the significance of the so-called Three-Age System cannot be overstated; by providing a chronological interpretation of archaeological finds, it laid the foundation of modern archaeology. But the Three-Age System also carried with it, almost as an undercurrent, the assumption that cultural progress is inevitable, and that it follows a series of stages that by their very nature lead one to another.

In *The Primitive Inhabitants of Scandinavia,* published in an English edition in 1868 (and excerpted in *The Origins and Growth of Archaeology*), the Swedish scientist Sven Nilsson explained:

> I feel more and more convinced that, as in nature we are
> unable rightly to conceive the importance of individual
> objects without possessing a distinct view of nature, so
> are we also unable properly to understand the signifi-
> cation of the antiquities of any individual country with-
> out at the same time clearly realizing the idea that they
> are the fragments of a progressive series of civilizations,
> and that the human race has always been, and still is,
> steadily advancing in civilizations.

Nilsson had made a leap: The advances in material technology from stone to copper to iron were merely symptoms. What they really reflected, Nilsson argued, was a
grand march of progress that included virtually every aspect of human life, and which followed a well-ordered
scheme through four distinct stages of "civilization." Since,
according to Nilsson's reasoning, all human cultures
throughout the world had gone through exactly the same
sequence of progress, he used contemporary hunter-
gatherer and nomadic cultures as the models for his early
stages. The first stage is that of the "savage," who acts only
for the "day that is, not the day which is coming" by
hunting and fishing—until "experience gradually awakens
reflection" and "the prudent thought then suggests itself to
him of saving a portion of his abundance of the day, and
still more, that of carrying away the young calf or fawn,
whose mother he has probably killed in the chase; and
collecting several more of them, and forming at last a herd,
he becomes a *herdsman*." The herdsman, upon abandoning
nomadism for a permanent fixed dwelling, becomes an
"agriculturist"; and finally, with the development of coined
money, a written language, and a division of labor among
professions, the fourth stage—that of a "nation"—is attained.

The sequence has since been proved incorrect; the cultivation of plants generally preceded the tending of stock. Good evidence for both domesticated grains and domesticated sheep and goats appear in the Near East about 9500 B.P.;* cattle appear sometime around 7000 B.P.; and horses, the last of the major domesticated animals, at about 5000 B.P. The dog is the only exception to this sequence, with its domestication predating the rise of agriculture by at least three thousand years.

But factual quibbles aside, Nilsson's chronology set the tone that would be adopted for years to come. The idea of inevitable cultural progress was carried one step further by Lewis Morgan, a Rochester, New York, lawyer whose hobby was studying American Indians. In 1877, he published *Ancient Society; or, Researches in the Lines of Human Progress from Savagery Through Barbarism to Civilization,* which explicitly put forth the idea that a sequence of seven cultural stages (such as "Lower Status of Savagery" and "Middle Status of Barbarism") was the same throughout the world, and that the stages are "connected with each other in a natural as well as necessary sequence of progress." An interesting and telling footnote is that Morgan's ideas were seized upon by both Marx and Engels as corroboration of their ideas that a sequence of cultural developments, leading ultimately to communism, is inevitable.

In all of these models, what was really nothing more than a factual description of an archaeological chronology (the Iron Age followed the Copper Age) was subtly trans-

---

* "Before the Present." Archaeologists usually find it easier to express time in number of years ago than to reckon with the switch from forward-running A.D. years to backward-running B.C.

formed into something that passed for an *explanation* of the chronology. Morgan proposed no underlying mechanism to explain how or why a culture passed from Lower Barbarism to Middle Barbarism or Stone Age to Copper Age; no mechanism was needed because, so it seemed, passing from one stage to another was natural and inevitable. The answer to the question, Why did a culture adopt agriculture? was, because that was the next rung on the ladder.

The more recent fashion in anthropology has been to look for reasons *why* a culture might change, but still to evade the question *how*. One leading current theory is that ancient societies adopted animal husbandry as a way to produce more food to feed a rapidly growing population; another sees it as a response to climatic or environmental changes that eliminated traditional sources of game.

Much of the debate takes place in the context of efforts to establish a theory of cultural evolution analogous to Darwin's theory of biological evolution. It is odd that even here, however, the older, pre-Darwinian ideas of progress still hold sway. Darwin's intellectual triumph was that he had explained the *how* of evolution, for the first time casting free from philosophical assumptions about the nature of progress. He produced an objective, scientific *mechanism*—natural selection—to explain what he had observed in nature. It is a two-step process: First, within each species there is a natural variation in traits. Second, those individuals whose particular traits make them more likely to survive and reproduce will pass them on to the next generation ("natural selection" of the "fittest" traits).

What Darwin was careful not to say was that evolution has any inherent goal or proceeds according to a plan. Although one consequence of variation and natural selection is that it can explain how species have adapted to their environment, it can equally well explain how species have become extinct, the exact opposite of "progress." In *The Origins of Agriculture,* David Rindos of the University of Western Australia cites the hypothetical example of a population of horses that is able to crop grass to within two inches of the ground. As the population begins to reach the limits of the food supply, those individuals that are able to crop the grass to within one inch of the ground (say, because of a different shape of teeth or mouth) will be fitter; they will be more likely to survive and pass on their traits, and eventually the entire population will be able to crop grass to one inch and the selection pressures again will favor those individuals that can crop the grass still lower. Finally, the entire population will consist of horses that can eat the grass down to the ground, at which point the grasslands are wiped out and all the horses starve to death.*

Still, the idea that evolution is synonymous with progress is still very much with us, however much Darwin may have tried to dispel the notion. Well into this century, popularizations of evolution persisted in portraying the evolution of species as an ascent up a ladder of life. The stair-step diagrams showing amoebas at the bottom, reptiles crawling onto land somewhere in the middle, hairy guys with clubs higher up still, and modern man, naturally,

---

* Darwin himself paid considerable attention to one large category of nonadaptive traits, the result of what he termed "sexual selection." The peacock's tail is the classic example: It cannot by any stretch of the imagination be called an adaptation to the environment; if anything, it is an encumbrance that requires the peacock to expend extra energy dragging it around and renders him more vulnerable to predators. But its value in attracting females ensures that those males possessing the trait will be "fit"—that is, they will be more likely to mate and pass on that trait—even though the trait is far from "adaptive."

at the top are so familiar that they have long been the stuff of parodies. (A cartoon by S. Harris has the heroic figure at the apex followed by an additional figure on the other side of the mountain, clearly the first in a descending series: a man watching television.)*

Rindos's point is that just as Darwin's theory could account for the extinction of species, so any genuine explanation for the domestication of animals should be able to account for the failure of some animals, such as the ancient Egyptians' gazelles, ever to be domesticated. The inability of the purely cultural explanations to do that reflects how much they still rely on assumptions about human progress rather than elucidation of an underlying mechanism of change.

Rindos also explains how the long-standing failure of many scientists to question the adequacy of purely cultural explanations for the domestication of animals is the result of a subtle rhetorical trap in the language of evolution. It is common, for example, for scientists to say that giraffes developed longer necks in order to eat the leaves at the tops of trees, or that as the climate became colder, bears grew thicker coats. In neither case does the speaker literally mean that there was an intention or goal behind the changes that occurred; he is summarizing in descriptive form the end result of a process whereby individuals with fitter traits (longer necks or heavier fur) were selected by

---

*Darwin was not the first to suggest that species evolve, and evolutionary theorists who preceded Darwin did explicitly equate evolution with progress. The French naturalist Jean-Baptiste Lamarck (1744–1829) wrote that he had originally developed his idea that species evolved through the inheritance of acquired characteristics as a pedagogic tool to explain to his students his overarching philosophy—that inherent in life itself is a vital force tending toward perfection.

environmental pressures (competition for food or climatic change). But it has the rhetorical form of a cause-and-effect explanation: It implies to the unwary that the *purpose* of the change, and not merely the end result of change, was adaptation to the environment. As Rindos shows in his example of the hypothetical horses selecting themselves to extinction, the only real "purpose" of evolution is to adopt *fit* traits, some of which may not be adaptive at all.

So when anthropologists propose the theory that *cultures* change in order to adapt to the environment, it takes the same rhetorical form of these familiar evolutionary "explanations." But in fact it is no explanation at all. It is as if a historian were to propose the theory that Abraham Lincoln became President in 1860 because it was an adaptive response to the nation's need for a strong leader, while another historian countered that economic forces were responsible, but it turned out that neither even knew, or bothered to ask, how Presidents are elected.

Rindos aptly calls the idea that domestication was a human invention "two parts rhetoric, one part truth, and one part wishful thinking." Part of what continues to cloud our thinking about the origins of domesticated animals is that in the recent past man has indeed directed the course of evolution of the animals under his control to a remarkable extent. The definition of dog breeds through intentional selective breeding toward predetermined goals, and in particular the development of exaggerated physical characteristics, such as the transformation of basically rounded heads that were the norm a hundred years ago in both cocker spaniels and Scotch collies to flat faces in cockers and chisel-shaped heads and elongated noses in collies, is a case in point. A glance at any seed catalog likewise shows how far we have gone toward mastery of a bewildering array of desirable characteristics in the breeding of plants, everything from a shade of color in eggplants to shelf life in pears. It is thus not surprising that our first inclination is to see domestication as a triumph of human power over nature; to an impressive degree, that is what it has become today.*

Yet the first agriculturists had to take what nature gave them. As Rindos argues:

> We ... forget that [man's] options in dealing with the biological world are in large part dictated by processes over which he has no control. Man may indeed select, but he cannot dictate the variation from which he must select. ... We might equally well describe the evolution of domesticated plants by saying that the plants chose humans to protect and disseminate them ... would we then have to discover ecological imbalances in the demographic structure of ancient plant populations that forced the plants to adapt this way?

The search for an explanation that can overcome the inadequacies of the purely cultural theories of domestica-

---

*It is always worth remembering what a recent development this is, however. Most breeds of dogs, for example, are nineteenth-century inventions. A 1576 treatise on British dogs by one Johannus Caius recognized only seventeen breeds, or really general varieties, most described by function ("Shepherd's Dog") or general type ("Water Spaniel"). And even today, with the new tools of genetic engineering, we cannot overcome some fundamental limits that nature imposes on our powers to manipulate organisms. Scientists have, for example, succeeded in inserting into plants a gene found in a bacterium, *Bacillus thuringiensis,* that produces a natural insecticide. But that hardly means it will ever be possible to create a half plant, half bacterium. By definition, new traits introduced into an organism cannot have more than a minor effect on the organism's overall character without derailing the development of the embryo.

tion must begin with biology. The archaeological record and the behavioral patterns of contemporary domesticated species have provided us with two clues that can point us toward an explanation.

The first, touched on above, is that the desirable characteristics that form the very definition of a domesticated species—such traits as docility, lack of fear, high reproductive rate, early sexual maturity, fast growth—were not to be found in the wild ancestors they descended from. Nor could the putative first farmer even have known that captive breeding would eventually produce such traits.

Even if some primitive man or woman did set out into the woods to bring home a wolf pup, even if a thousand of them did, that act of human initiative cannot explain the vast biological differences between wolf and dog. Many people do raise raccoons, skunks, and wolves in human households, but such tame animals are far from domesticated. They show a degree of aggressiveness and unpredictability that sets them well apart from the behavior of dogs and cats. Sheepdog handler John Holmes neatly demonstrates this with a photograph in his book *The Farmer's Dog,* showing a Border collie and a tame fox both eyeing a tame white rabbit. Both look like they are about to pounce, and indeed the fox clearly would have already if not for the restraining leash. The dog, however, is checked in place only by a voice command. Its hunting instinct is attenuated to the point that training alone can keep it from following through even the most compelling natural instincts of its wild progenitors.

Domesticated animals display many other behavioral characteristics, in particular a loss of defensive, fearful, and territorial behaviors, that reflect basic biological changes, not just training or taming. One anecdotal but nice example can be found in the tracks of dogs and cats in a new fallen snow. The tracks of wild animals such as wolves and foxes move in straight and purposeful lines; at the trot, each hind foot is cautiously placed exactly in the spot where the corresponding front foot first fell. Dog and cat tracks are a testimony to carefree fearlessness. They meander, circle, zigzag, and drag; the hind feet sloppily miss the mark.

The basic biological changes that underlie this loss of fearfulness have been well documented in that most studied of animals, the laboratory rat. A domesticated version of the Norway rat *Rattus norvegicus*, the laboratory rat has undergone substantial changes in its major hormone-producing glands. The adrenal glands, associated with the hormones that trigger the fight-or-flight response, are much smaller than those of its wild counterpart. Laboratory rats show little or none of the fear of strangers, loud noises, or new situations so marked in wild rats.

The obstacles that a Mesolithic pioneer would have faced in restraining in captivity members of a wild species— restraining them for the many generations it would take to systematically breed the domesticated traits we see today— are enormous. The behavior of even tamed wild animals often becomes disruptive once they reach sexual maturity, and they will try to escape and return to the wild. Attacks by tame wolves have been reported when members of the household behave in a way that the wolf interprets as prey behavior—such as a child who cries, stumbles, or runs away. To bring a young animal in from the wild and keep it around as a pet, as the aborigines do, until it starves to death or runs off, is one thing; systematic captive breeding of rambunctious or even dangerous animals is quite another. Transforming the wild to the domesticated is a far greater challenge, even in strictly genetic terms, than are such modern-day accomplishments of captive breeding as transforming a purple eggplant into a white one, or a short collie nose into a long one.

The second clue that may help guide us to an understanding of how domestication happened is that the consequences of domestication at times not only have nothing to do with human intention, but in some cases actually work in opposition to human intention. Even now, after millennia of domestication, it is sometimes difficult to discover any of the supposed benefits that would have prompted the first farmer to go to this trouble, or that prompt people today to continue to care for and feed these animals.

Much of the mythology surrounding the dog, for example, focuses on his utility as a guardian or hunter or herder. That is always assumed to be the reason humans tamed and domesticated the dog. Certainly it is true that these are benefits man has gained from his association with dogs. One can also argue that, in Western society at least, dogs and cats serve mankind by nothing more than the companionship they offer.

But the success of dogs and cats in association with human society far outstrips any of these benefits to humans, even when the benefits are very broadly defined to include companionship. As much as we in the United States lavish care and affection and money on our pets, the reproductive success of the country's one hundred million dogs and cats outruns our generosity, witness the millions of unwanted cats and dogs that are killed each year in animal shelters. In most parts of the world, cats and dogs are looked on much less favorably; but even where they are considered vermin and are the targets of occasional extermination campaigns, they continue to flourish.

Their survival has nothing to do with being rewarded for their utility to man. It has to do rather with their superb adaptation to human society. In many tribal societies, for example, dogs live in loose association with humans. They are not captively bred; there is no concerted effort whatever to select for any desirable characteristics. Yet they are clearly domesticated—that is, they are not wild dogs that have been brought up from pups and kept as pets in the mode of the Indians' raccoons. In the case of the Beng people of Africa, the dogs appear to serve no useful purpose whatever—they are not working animals, they are not even eaten. They are not pets, either. They are tolerated but are shown no affection. Biologist Raymond Coppinger observed a similar phenomenon in villages in the Andes, although there the villagers actually tried to run off the dogs, which on balance were considered a nuisance, biting people and spreading disease. Coppinger describes the relationship of dog to villager as one of social parasitism. The dogs' submissive behaviors—such as cringing, looking away, or rolling over—are readily recognizable as such to humans, and generally succeed in deflecting attempts to chase the dogs off, or worse.

Even the domestication of food animals and the rise of animal husbandry, which conventional wisdom has always portrayed as a revolutionary advance for mankind, now appears to have been something that no sane hunter-gatherer would have adopted by free choice. However much our modern society may rest upon agriculture, there is a mounting body of archaeological evidence that indicates that the transition from foraging and hunting to growing corps and tending herds was, as physiologist Jared Diamond puts it, a "catastrophe."

Reviewing the data, Diamond only slightly tongue in cheek calls the adoption of agriculture "the worst mistake in the history of the human race." Paleopathological studies show that, at least initially, agriculture triggered an epidemic of injuries, malnutrition, and infectious disease. Studies of modern hunter-gatherers, such as the Kung of the Kalahari Desert of southern Africa, show that their daily intake of calories and protein is well above the recommended daily allowances; the average time per week spent gathering food for one group of Kung was only twelve to nineteen hours. The bones of ancient agriculturists tell a very different tale. For one thing, farming appears to have introduced an array of new occupational hazards: Skeletons of the human residents of Tel Abu Hureyra in northern Syria show deformities in the legs, feet, and vertebrae, probably caused by the stresses of carrying heavy loads.

Nutritional problems and susceptibility to infectious disease seem to have been an even worse consequence. Anthropologist George Armelagos studied the skeletons of Indians who lived in Illinois from A.D. 950 to A.D. 1300; their adoption of intensive agriculture in A.D. 1200 was accompanied by a sudden increase in disease. In the preagricultural phase, only 16 percent of the skeletons showed signs of iron-deficiency anemia. After A.D. 1200, the incidence shot up to 64 percent. The overall rate of infectious diseases that leave a mark in bone went from 27 percent to 81 percent. Average life expectancy dropped from twenty-six to nineteen years.

Unmistakable signs of osteoporosis—in modern times considered primarily a disease of old age, since most patients are women over fifty—appear in the bones of women of childbearing age in an agricultural community in Nubia five thousand years ago, apparently a sign of extreme calcium deficiency in the diet.

The explanation for all of these signs of nutritional deficiency is fairly clear: Hunter-gatherers have an extremely diverse source of food that not only results in a nutritionally more balanced diet but is also less susceptible to drought or other environmental stresses. Agriculturists are often dependent on a single vegetable crop for their survival, as even such relatively recent catastrophes as the Irish potato famine illustrate. And even in the best of times, a single, starchy vegetable is far from a complete source of necessary proteins, vitamins, and minerals.

Direct evidence that the agricultural way of life was not such an obvious improvement is buttressed indirectly by evidence that shows agriculture being adopted slowly—one might even say reluctantly—rather than being embraced in an enthusiastic "revolution." The picture of an agricultural revolution first started to unravel when digs in the Near East showed the adoption of farming to have been a gradual process that went on side by side with foraging and hunting for thousands of years. Tel Abu Hureyra's story appears to have been typical. The village was occupied continuously from 11,000 to 7000 B.P., with one five-hundred-year break beginning about 10,000 B.P. An analysis of myriad artifacts and plant and animal remains, including some sixty-thousand animal bone fragments, recovered from the site before it was flooded by the Tabqa Dam project in 1974, tells a story of anything but a revolution. For the first fifteen hundred years of occupation, the residents of Abu Hureyra continued hunting gazelles and gathering wild plants, with no evidence of agriculture. Even after domesticated grains appeared, about 9500 B.P., hunting continued along with agriculture for at least another thousand years, as evidenced by substantial quantities of gazelle bones. Radiocarbon dating of bones shows that up until 8500 B.P., gazelles accounted for 80 percent of the bones found at Abu Hureyra, sheep and goats only 10 percent.

It is unclear whether the sheep and goats were wild or domesticated in the earliest phase. But by around 8500 B.P., clear evidence emerges that sheep and goats were indeed domesticated—even as gazelle hunting continued to supply the overwhelming preponderance of meat for the dwellers of Abu Hureyra. The conclusion is the result of some clever analyses of the animals' teeth. Gazelles are born with three baby molars, each with a prominent cusp at the center. Nearly all the young are born in April and May. During the course of their first year, as they put their teeth to use eating tough grasses and plants, the cusps of these so-called milk molars are steadily worn down. In late summer of the following year, they lose their milk teeth, which are replaced with permanent premolars. Archaeologists Anthony Legge of the University of London and Peter Rowley-Conwy of Cambridge measured the height of the cusps on the milk teeth they found and discovered that nearly all the teeth fell into two groups: those with almost no wear, corresponding to newborns, and those with substantial wear that corresponds to animals just about one year old. The clear implication is that hunting was highly seasonal. All the animals were killed in the late spring, apparently as they arrived in the vicinity of Abu Hureyra at the end of their northward spring migration.

There were too few sheep and goat teeth dating from the early period of settlement to form any clear conclusion about the status of these animals. But teeth from around 8500 B.P. show a very different pattern from that of the gazelles. Instead of falling into distinct groups, the teeth show a continual gradation of wear, indicating that the animals were killed all year round—suggesting either domestication or at the very least an incipient domesticated relationship wherein sheep and goats coexisted with human settlement on a permanent basis. In other words, even after domestication of grains and probably after domestication of animals, the citizens of Tel Abu Hureyra continued to rely upon hunting for their primary source of food. During this period, the residents built substantial mud brick houses and began manufacturing stone vessels, jewelry, and plaster. While still carrying on a "primitive" means of subsistence, they had adopted an increasingly "civilized" way of life.

Archaeological evidence from Europe likewise tells a story of slow, sporadic adoption of agriculture—just the opposite of what would be expected if it were an invention of revolutionary import. In northern Germany, farming was being practiced by 6500 B.P.; to travel the relatively short distance to Denmark and southern Sweden took at least twelve hundred more years. In some parts of eastern Europe, an agricultural way of life was not adopted until 4500 B.P. or even later, even though some knowledge of farming, manifested by small numbers of bones of domesticated animals, had clearly reached the area. Europe in this period was a patchwork. Foragers persisted in parallel with agricultural communities for thousands of years. Their artifacts show that the foragers often developed highly specialized tools, such as harpoons for hunting seals in the case of the so-called Pitted Ware culture that lived on the coasts of Sweden about five thousand years ago.

And, as at Abu Hureyra, hunting and gathering was often able to provide a level of prosperity sufficient to support a "civilized" mode of existence, complete with permanent settlements, a social hierarchy (implied by a differentiation in graves; certain members of the community were buried with large numbers of bone spearpoints), and the manufacture of animal carvings of wood, stone, and bone. The obvious conclusion is that the advantages of agriculture were far from overwhelming.

The implications of all of this evidence is that if we are to make sense of how and why domestication happened— happened in spite of good reasons why humans should never have gone along with it, and in spite of biological obstacles that would have made it almost impossible even for humans determined to ignore good reasons—we need to stop looking at domestication as only a human phenomenon. The direction to start looking is through the eyes of animals. The question that must be answered is not what's in it for us, but what's in it for them.

III

THE
VIRTUES OF
DEFENSELESSNESS

If you pick up a starving dog and make
him prosperous, he will not bite you. This
is the principal difference between a dog and a man.
—MARK TWAIN, Pudd'nhead Wilson

Many animals have found it in their interest to associate with humans. We've just been slow to recognize the fact.

"Must we kill the mouse?" a little girl asks her mother in an advertisement in *The Animals' Agenda* (subtitle: *The Animal Rights Magazine*): "No, honey," the mother replies, "the world is big enough for all of us. The mouse is intelligent and wants very much to live. We just want him out of our home and into the woods where he belongs." The solution—available for a mere ten dollars—is a live trap complete with patented "Freedom Door" ("delays escape so mouse must eat thru [sic] cracker to freedom, avoiding contact with captor," though it is not made clear which party this feature is meant to benefit).

The significant feature of this advertisement, however, is the contention that what the mouse really wants is to be out in the woods—back in "nature." It is a view that fits well with the simple picture of nature as a place where animals live out their peaceful lives, far from the disturbing influence of man. In Takoma Park, Maryland, which has the distinction of being a nuclear-free zone, one of the handful of American cities that can boast of having elected a socialist mayor, and home to a number of vocal animal-rights activists, this nice picture of an orderly natural world came into play on a slightly grander scale a few years ago when the town was overrun with rats. The local government decided to provide residents with traps of the traditional, lethal variety. So the animal activists swung into action, and demanded that live traps be offered instead to those who wanted them. Coming to the defense of an animal usually considered a public-health menace was at least ideologically laudable, as the animal rights movement, in spite of its philosophical credo of equal rights for all species, tends to be partial to the rights of cute, furry animals readily amenable to anthropomorphic projections. But the problem then arose what to do with a large collection of captured rats in what is a very urban suburb of Washington, D.C. The animal activists suggested that rats could then be released into a "natural" habitat. Where? Frederick County, suggested one. The farmers of that rural county, who are among those who bear the brunt of an estimated one billion dollars in damage caused annually by mice and rats, were not amused.

But the mice and rats in question would probably not have been amused, either. The image of furry animals safely nestling in the woods has little to do with the evolutionary reality of rats and mice. After eating its way through the Freedom Door, the trapped mouse will either return to a nearby house, be eaten, or freeze or starve to death in the woods. Neither rats nor house mice are native to North America or Europe. To be sure, the house mouse *(Mus musculus)* will on occasion nest in meadows or grainfields (not woods); in temperate areas such as California and Florida the black rat *(Rattus rattus)* will nest in trees; and the Norway rat *(Rattus norvegicus)* will summer in grainfields. But all live in close association with man by preference. By virtue of several thousand years of shared evolution and an ability to eat anything that we do, rats and house mice are supremely adapted to living inside walls and garbage dumps. Biologists call the species "human commensals"—they eat at our table, sometimes literally. The warmth, incidental protection from predators (we like sharing our shelter with snakes even less than with mice and rats), and bountiful food supply of a house provided a niche that allowed rat and house mouse populations to explode and spread over the face of the earth. House mice, in fact, compete poorly with most wild indigenous small mammals; their success is a direct result of their ability to colonize human habitations. The black rat and Norway rat, which originated in the Far East, likewise became human commensals as soon as the first permanent human settlements were established. The opening of trade routes allowed the black rat to reach Europe in the Middle

Ages (where it so ably assisted the spread of the bubonic plague); traveling aboard ships it arrived in America in the sixteenth century and quickly became established in the towns along the East Coast. It is now found on both East and West coasts. The only parts of the world where the black rat flourishes in the "wild" is in woody areas in the warm climates of Southeast Asia, its presumed region of geographic origin. Black rats have colonized most of the islands in Southeast Asia, though even there man probably played a part, inadvertently carrying the rats aboard ships. The Norway rat, denizen of sewers, basements, and dumpsters, spread from Asia in the early eighteenth century into Europe, and arrived in America around the time of the American Revolution. It is now found in every state of the Union—wherever people have concentrated.

So the solution is not so neat and simple as one of restoring the mouse to "nature." When it's in our house it already *is* in nature. By its evolutionary history, by its feeding habits, and by its nesting habits, cohabitation with man is the way of life the house mouse has adapted to.

The evolutionary scheme that led rats and mice to drop their natural defenses and approach human habitation as a way of life is one that has been employed many times in the history of this planet. A close look at other evolutionary odd couples in nature—species that have gained an advantage in the struggle for survival by flocking with another species in defiance of the norms of defensive behavior—reveals biological motives that are apparent in all domesticated relationships. The argument that domestication is such a departure from the natural order and so sophisticated a development that it could only have been the result of human volition is answered by the craftsmanship of nature herself.

Those animals that have forged cooperative relationships with man that fall short of true domestication are especially revealing, both about biology and about our assumptions of what is natural. Among the many fantastic tales that European explorers brought back from what novelist Laurens van der Post so aptly called the "ancient Merlinesque world of Africa" was that of a bird that would lead men in search of honey to bees' nests hidden in tall trees on cliffside crevices. Travelers' tales being what they are, especially travelers' tales of exotic animals, the story of the "honey guide" bird, which first appeared in print three centuries ago, was long dismissed as a myth. With the tide of modern agriculture and commercial beekeeping sweeping aside the traditional ways of African tribesmen, there the story might well have rested: a story, no more, no less, just one of thousands of ancient, Merlinesque tales of wizards who turn themselves into lions to hunt for meat and of crocodiles that employ magical powers to make men drop their guard.

But happily for science, the nomadic Boran people of northern Kenya still collect wild honey following the practices of their ancestors, and more happily still, one of their number, H. A. Isack, became an ornithologist at the National Museum of Kenya. Isack decided to conduct a rigorous scientific test of his fellow tribesmen's claims—that the honey guide *(Indicator indicator),* by its flight pattern and vocalizations, tells the honey gatherers the direction and distance to a nest, guides them to it, and alerts them when the nest is reached.

In a three-year field study, Isack found that, far from a myth, the behavior of the honey guide is a remarkable example of social cooperation between species that is purely the result of natural forces of evolution. Both parties benefit. The Borans save many hours spent in futile search for nests: Isack found that in unfamiliar terrain, honey gatherers spent an average of nine hours to locate a nest on their own, versus three hours when they followed a honey guide. The birds, in turn, obtain a source of food—bee larvae and wax—accessible only after the nests are cut open by humans.

In a typical encounter, a honey guide attracts the attention of the Borans by swooping down, flitting from branch to branch, and loudly and persistently calling; it then takes off and flies above the treetops in the direction of a bee's nest that can be more than a mile away. After anywhere from a quarter of a minute to several minutes the honey guide will return, again calling, and proceed to lead the humans in a very nearly direct route to the nest, stopping every hundred feet or so for its charges to catch up. Upon reaching its goal, the honey guide will perch near the nest, emitting a distinctive "indication" call in place of its guiding call, and will sometimes circle the nest.

The Borans, for their part, have developed methods for summoning a honey guide with a distinctive whistle produced by blowing into their hands or into whistles carved from snail shells or palm nuts. In one experiment, the researchers summoned the same bird to seven different starting points around the entire compass; each time, the bird led them on a near-direct route to the same nest.

In trying to understand the phenomenon of domestication, the honey guide is not a bad place to start. The case of the honey guide forces us to face our assumptions about what is "natural" behavior and what is "man-made." And it points up some general rules about why certain animals in nature have, as part of their evolutionary strategy for survival, adopted the practice of living cooperatively with members of an entirely different species. In the case of the honey guide, a naive observer (one not so sophisticated as to dismiss it all as a myth, that is, yet still sophisticated enough to reject enchantment or creation by design as possible explanations) might very naturally assume that the birds had been tamed and trained for their task, much as falcons are for hunting or cormorants are for fishing. The honey guide's behavior certainly is at odds with what seems natural for a wild creature. It shows no fear of man; rather than camouflaging itself, or taking flight at the approach of humans, it actively seeks them out, calling and flashing white tail feathers to get their attention. In one sense, it thus is "tame."

But did humans tame it? Not in any normal sense of the word. It is a free-living animal, and always has been. Humans do not capture honey guides and rear them in confinement; they do not breed them captively; they do not train them as one trains a dog to lie down or a horse to canter on command.

Yet, no matter how "natural" the honey guide's behavior is, it is a product of forces that clearly include man's behaviors—man's ability to use tools to cut open beehives; his fondness for honey, though not for bee larvae; and his conscious awareness of the behaviors of the animals around him. These human characteristics formed part of the evolutionary template that has guided the selection of tame and cooperative behavior among the honey guides. Humans did not deliberately set out to tame honey guides and train them to be guides to honey nests; nonetheless, conscious human action, unconscious though it must have been of the end result, did steer the evolution of this symbiotic relationship.

The distinction is a crucial one. To suggest that domestication is an evolutionary phenomenon rather than a human invention is not to claim that humans were mere pawns in some grand preordained plan. Human choices and human ingenuity clearly must have played a part—but only a part, insufficient in itself. It goes back to David Rindos's observation that in dealing with the biological world, humans may select, but only from a set of options determined by forces beyond their control.

Even today, in this age of human technological mastery, there are many examples of "wild" animals that are adapting to human society in ways that underscore the limits of human control. Norway and black rats and house mice are just a start. The barn swallow (*Hirundo rustica*), for example, is called that for a very good reason: It actually prefers to make its nest in barns, gaining not only physical shelter but also some likely protection from predators as a result of the proximity to humans. Any country dweller in the United States knows how well adapted starlings are to human dwellings; they are almost impossible to get rid of, entering through chimneys, small holes in siding, and attic vents.

Many birds have adapted even to modern man's technological habits. Sea gulls mobbing fishing boats to steal part of the catch or to scavenge refuse thrown overboard are such a familiar sight that we take them for granted, a fixture of seacoast life. But it is worth looking at this adaptation in light of such examples as the honey guide. The gulls actually pursue boats for long distances. It is more than just an opportunistic encounter; it is a way of life for gulls. To the extent that they rely upon food from fishing boats (and, apparently, upon the fish that, disturbed by the boats' passing, rise to the surface, making an easy catch for the birds), gulls have actually become dependent upon human society. Many other species of birds have become similarly adapted to agricultural practices, following tractors to feed upon worms and insects flushed by the machinery and the plow. Falcons will even pursue trains, to prey upon the small birds flushed from along the tracks.

The propensity of raccoons to raid human garbage cans, deer and rabbits to feast on gardens and cultivated trees, birds (and squirrels) to feed at bird feeders, and even bears to feed on handouts from tourists in parks are similar contemporary examples of animals that have adapted well to life on the fringes of human society.

The mutual adaptation of the Laplanders and the reindeer (*Rangifer tarandus*) herds they follow is perhaps the closest example to domestication of any of the "obviously" natural interactions between humans and animals. Both the Lapps and the reindeer herds are nomadic. Throughout most of their continual migrations, the reindeer run free. The Lapps hunt the reindeer, yet the relationship is very different from that of predator to prey; as the zoologist Frederick Zeuner long ago pointed out, it is at times difficult to tell who is following whom. As Zeuner explains, the free-living reindeer are bound to the human campsites just as the humans are bound to the reindeer herds:

> Human urine is regarded by [the reindeer] as a great delicacy, and it is this substance which attracts and binds reindeer to human camps. This craving is probably due to the lack of salt in water available to reindeer, which is mainly derived from melting snow, though all ruminants are attracted to salt-licks. The reindeer nomads of course take full advantage of it, even today, so that the supply of a delicacy provides the meeting ground on which the social media of the two species "overlap."

Who tamed whom? The answer is far from obvious. Although in recent times some reindeer have been fully domesticated, and used as dairy, draft, and saddle animals, this development came only after centuries of a more equal symbiosis, which continues. In one sense, humans are social parasites upon the reindeer: They mimic the social behavior of the reindeer (nomadism) to their own advantage. Yet the reindeer remain largely free-living animals that, as it were, decided of their own free will to associate with humans. Their behavior is certainly very different from that of, say, white-tailed deer or elk in North America, which, when approached by humans, show the typical flight response of prey encountering a predator. There is something going on here that is more than just a trick of luring wild animals with a baited trap; the reindeer's association with humans, and their relative tameness, reflect some underlying biological changes that have occurred over the course of centuries of evolution alongside humans.

The ways that species throughout nature evolve toward cooperative behavior share a number of characteristics with domestication. Whether one considers reindeer and sea gulls or dogs and sheep, all have exchanged some "wild" attributes—the characteristics traditionally associated with majestic, independent creatures, such as fear, aggressiveness, territoriality—for some tangible advantage, usually either protection or food, that confers increased survival value.

Biologists broadly apply the term *coevolution* to the process whereby two species evolve in concert. It is not hard to see how one species' behavior can influence the evolution of another's. The environment that a species inhabits often includes the behaviors of other species, which thus become a force in determining its evolution. This is readily apparent in the case of a predator-prey relationship. The behavior of a lion in preying upon antelope creates a selective pressure that favors those antelope with more effective defense mechanisms, such as an ability to jump and run quickly. In turn, the behavior of the prey creates selective pressures that favor certain characteristics in the predator, such as sharp claws, a keen sense of smell, or an ability to stalk. The myriad array of defense mechanisms that have evolved throughout nature, from the camouflaged coloring of the snowshoe hare and the sting of the wasp to the complex antibiotic chemicals exuded by fungi, is profound testimony to the creative resources of evolution in developing adaptations to other creatures' behaviors.

Predator-prey relationships are the classic examples that we think of when we consider evolution—which, after all, is supposed to be about the survival of the fittest and redness in tooth and claw. But coevolution has also produced an astonishing variety of mutualistic behaviors, some of which frankly surpass, in their seeming ingenuity, anything found in the relationship between man and domesticated animals. To take just one example, consider the behavior of the fish known as the cleaner wrasse (*Labroides dimidiatus*) and the large "customer" fish it serves. Parasites are rampant in the marine environment, and the fins, gills, and even mouths of many large fish are home to countless small parasites. The cleaner wrasse helps both itself and its customers by eating these parasites. To avoid being attacked when it approaches its customer, the cleaner fish signals with a distinctive "cleaner dance." The customer then spreads its fins and gill covers and opens its mouth as the cleaner bumps each with its snout, and holds each part still while the cleaner does its job, meticulously vacuuming off the parasites.

By drawing on this and the many other examples of coevolved relationships in nature, we can hope to discover what the basic biological motives are that drive animals to cooperate at all—and so better understand exactly how and why domestication might have arisen.

Biologists who have studied cooperative behavior have uncovered at least a half-dozen specific motives that explain why different species should flock together. One of the most basic motives is to pool defensive resources. Simply put, members of a flock benefit because many eyes are better than two. That is one reason why birds, horses, fish, and countless other species congregate in large single-species groups. In mixed-species flocks, such as the herds of giraffe, zebra, and wildebeest that are always seen grazing on African savannas in picture postcards and wildlife documentaries, the members of the group gain an added advantage because the especially acute senses of one species can make up for the deficiencies of another. Zebras are nearsighted, but have an excellent sense of hearing. Giraffes are farsighted. Pooling their resources gives them a greater chance of detecting an approaching lion than any one would have on its own.

Some interspecies flocks or associations appear to be the result of one species' taking advantage of the active defenses incidentally afforded by the presence of another species. Barn swallows and house mice probably fall into this category in their relationship with man. Similarly, the yellow-rumped cacique *(Cacicus cela),* a finch that lives in Central America, often builds its nests near wasp nests. The wasps afford considerable protection against the botfly, a parasite that lays its eggs on newborn chicks, and whose larvae then burrow in and feed on the flesh of the young birds. In one study, the caciques that built their nests close to wasp nests were found to be virtually free of bot-flies and reared nearly three times as many hatchlings as did those that nested away from wasps. Some species of fish will lay their eggs at the base of an anemone, whose sting-ing tentacles afford a similar measure of protection.

Such relationships don't necessarily require a quid pro quo. It doesn't cost the wasp anything to have a nest of caciques nearby or the anemone to have fish eggs by its side. Sometimes, however, such interactions can lead to a more highly developed symbiosis. Many species of crab, for example, carry anemones on their shells; the crab gains protection and the anemone gets bits of food that float by as the crab tears its prey apart. In some of these symbiotic relationships the anemone in fact takes the initiative, de-liberately seeking out shells occupied by a hermit crab. In other cases the crabs will take the lead, finding an anemone and carefully placing it on its shell. In these cases, though, the anemone still cooperates. It apparently can recognize the touch of the crab's claws and responds by releasing its grip on the rock or empty shell it was occupying to allow the transplantation to proceed.

Even in the absence of such a profitable pooling of skills, there are simple mathematical advantages that can explain both single-species and mixed-species flocks. One is the convoy principle. When German submarines began sink-ing Atlantic shipping in World War I, it quickly became clear that because the oceans are very large and ships are hard to see at a distance, one very easy way to cut losses was to send ships together in convoys rather than singly. A group of ten ships is only slightly more visible to a U-boat than is a single ship, so grouping them together reduces from ten to one the number of opportunities that a ran-domly patrolling U-boat would have of encountering a ship. Even if a submarine does locate a convoy, it is not likely to succeed in sinking all ten ships before being driven off by depth charges or being eluded by evasive action.

The case for flocking is even stronger if the predator kills only one member of a herd during any given hunt, as in the case with wolves, lions, and hyenas. That would be analogous to a convoy facing U-boats armed with only one torpedo apiece. For any individual ship, the chance of being discovered would be the same (or only slightly greater) if it was in a convoy than if it was on its own, but its chance of being killed when it *was* found would be one tenth as great (in the case of a convoy of ten ships).

Flocking also offers the possibility of employing a defense analogous to dropping depth charges. Depth charges are not terribly effective weapons against submarines, which can sneak up from any direction without warning; a single ship facing a single submarine is at a decided disadvantage. Similarly, a single musk-ox would find its horns of little value in fending off an attack by a wolf. However, carpeting the seas with depth charges can be quite effective in repelling an attacker—as is the defensive circle, with horns lowered pointing outward, that musk-oxen form when faced with a wolf attack. Another instance of this sort of collective security in action is the not uncommon sight of a cat disgustedly slinking out of the woods, pursued by a mob of squawking and dive-bombing birds. A single bird is usually no match for a cat's stealth and agility; a flock of birds can neutralize any cat's best-laid plans.

A related advantage is what is known as the confusion effect, which has been well documented in the case of predators that attack schools of fish. A predator confronted with many targets can find it difficult even to decide which one to go after, like the fabled donkey that starved to death, unable to decide which of two piles of hay to eat first. Even when a predator does make up his mind, he may be distracted by the sight of many different moving targets, which makes it difficult to track any one. (Predators, interestingly, often solve the problem by going after the one member of a flock or herd that acts or looks different from the rest. Hawks, for example, will preferentially attack a single white pigeon in a black flock or a single black one in a white flock.)

An advantage in numbers likewise falls to flocks of predators. Many eyes are more likely to spot prey; so long as the members of the pack remain within sight or voice of one another, they can also, to the increased benefit of all, fan out to cover a much wider area. Even members of different species, competing for the same food sources, can under certain circumstances all benefit from hunting together. Insect-eating birds of various species will often flock together when food sources are found in spotty patches. Although there is less food to go around when a source of food is found, the group is more likely to find food and to avoid wasting time searching areas already picked clean by others. Studies of mixed flocks of foraging birds show that the per capita time spent searching for food decreases when the animals move in groups.

Another motive behind interspecies flocking is apparent in those instances where one species uses another as a "beater" to flush prey—as in the examples cited above of falcons following trains or birds following tractors. The cattle egret *(Bubulcus ibis)*, for example, follows elephants and buffalo, feeding off the insects flushed by the large animals. In North America, the brown-headed cowbird *(Molothrus ater)* developed a similar association with bison,

but with the dwindling of the bison herd has now transferred it to cattle. In these cases, as with some of the mixed
flocks of prey animals, there is no quid pro quo between
species; neither is there competition, however. One gains
an advantage merely by associating with another; the latter
is totally indifferent to the former's presence.

Finally, predators will sometimes associate with members of another species with a parasitic aim. Hyenas will
follow packs of the African wild dog *(Lycaon pictus)* to
steal the prey killed by the dogs. In this sense they act more
like muggers than associates.

A particularly important lesson that students of domestication should heed from natural examples of coevolution
is that the motive is not always obvious. Even in relationships that seem to be purely parasitic or predatory, the
advantages do not always fall as one-sidedly as they might
at first seem to—a cautionary tale for those who glance at
a cow being led to slaughter and wonder what's in it for
the cow. One of the strangest examples in the natural
world of parasitism with a silver lining is the so-called
brood parasitism practiced by the giant cowbird *(Scaphid-
ura oryzivora)* of Central America. The cowbird has developed a remarkable system for avoiding the considerable
investment of labor entailed in building a nest and brood-
ing and rearing young: It sneaks its eggs into the nests of
other birds and lets them do the work. The host species,
which include the cacique and similar birds, nest in huge
colonies. Most cowbirds lay eggs specifically "designed" to
mimic those of the host; the female will search for a nest
that already contains several eggs laid by the host, and
attempt to slip in unseen and add a single egg of its own to
the clutch. The host species are, however, adept at spotting
mismatched eggs and throwing them out.

But astonishingly, the cowbirds that parasitize certain cacique colonies lay eggs that look nothing like the host's, and these cowbirds make no effort whatsoever to conceal their intrusion; they will enter a nest and lay their egg in the full presence of the caciques. Even more oddly, the hosts make no effort to discriminate between their own eggs and even obviously mismatched ones.

What is going on here? The explanation comes back to the other parasite that troubles the cacique—the botfly. It is only in those cacique colonies constructed outside the protective umbrella afforded by a nearby wasp's nest that the parasitizing cowbirds abandon stealth. And they can afford to do so—and are tolerated—apparently because the presence of a cowbird has the same antibotfly effect of the wasps. In those nests containing a cowbird, all the chicks remain free of botflies, even though they are not protected by the nearby wasps. Apparently the cowbird chicks are able to repel the botflies, accomplishing the same thing. Feeding an extra mouth is apparently a small price for the host to pay to ensure the survival of its own chicks. Those members of the host species that, under these circumstances, tolerate the cowbirds' presence are more likely to produce viable offspring; because they tolerate the cowbirds, the cowbirds that choose to lay their eggs in these nests do not need to invest any extra energy into producing mimetic eggs or adopting furtive behavior. A single benefit is enough to transform parasitism into symbiosis, to recast a relationship in which one partner is exploited to one in which that partner is a willing participant.

The more one understands the motives behind coevolution in the wild, the less one feels the need to invoke the deus ex machina of human invention to explain domestication. Dogs, for example, likely benefited from their association with humans in a number of ways. They would have gained immediately from scavenging human camp-sites and parasitizing human hunting parties, much as hyenas parasitize the African wild dogs. Feral dogs—that is, domestic dogs that have been abandoned as pets or allowed to run freely in wild packs—studied on a Navajo Indian reservation in the American Southwest, for example, were found to depend heavily on open garage dumps for their food. It is not difficult to imagine free-living bands of humans and wolves engaging in more cooperative hunting, either; dogs and humans could flush prey for one another and use their different skills to obtain game that neither could on their own. Dogs have a keen sense of smell and fast legs; people had spears and bows and arrows. Hunting with a bird dog is a quick lesson in how the skills of the two species complement one another. Dogs can track, spot, and flush game that a person would never even see. Some primitive hunters today use dogs to considerable advantage as a sort of herd dog; the dogs actually hold the game at bay until the hunters can draw close and kill the animal with weapons that have either too short a range (spears or bow and arrow) or that take too long to set up and fire (matchlock rifles) to be used effectively unaided. Together, they can bring down much more game than either could on his own.

With nature as our guide, it is not hard to discover the biological motive that would have led even animals that our ancestors ate, such as sheep or cattle, into closer association with humans. Those animals would have gained measurable advantages from flocking with humans—being able to scavenge campsites or grainfields and live under a shield that guarded them from other predators.

One might wonder whether the advantages of extra

food and protection from other predators would outweigh the disadvantages of being preyed on by humans. From an evolutionary point of view, the answer is undeniable: A handful of minor species emerged from scraping together a marginal living at the end of the Ice Age to occupy a position of overwhelming dominance in the biosphere currently account for about 20 percent of the total biomass. Domestic dogs, sheep, goats, cattle, and horses far outnumber their wild counterparts. The global populations of sheep and cattle today each exceed one billion; their wild counterparts teeter on the brink of extinction.

Horses, in fact, would very likely be extinct today had it not been for their domestication. In North America, the once abundant herds that roamed the continent during the last Ice Age, some fifteen thousand years ago, had vanished by 8000 B.P., the victim of a rapidly changing climate that swept forests northward into the once open plains, and of hunting by humans. (A related species, *Equus hydruntinus*, that lived in southern Europe and western Asia during the same period was also wiped out completely.) In Europe and Asia, the range of the wild horses shrank and shrank in the postglacial period as grasslands disappeared and the herds moved farther and farther east; in modern times, the only remnants of the original wild horse herds (as opposed to the so-called wild horses of the American West, which are actually descendants of domesticated horses brought to the New World by Columbus and other Europeans) were the Przewalski's horses in the Mongolian steppes, which today survive only in zoos.

The process was dramatically reversed around 5500 B.P. According to radiocarbon dating of bones found at a site in the southern Ukraine, that was the period when the horse entered domestication. Horses had been hunted for thousands of years earlier, so the mere presence of horse bones is not confirmation of domestication. Several clues at the Ukraine site, known as Dereivka, however, point strongly to a change from prey to a very different status: Some 60 percent of the animal bones at the site were horses, and most tellingly, six bridle bits made of antler were found as well. From there, the domestic horse began rapidly repopulating the territory that a changing climate had driven it from. Shortly after the first signs of domestication appear in the archaeological record, horses begin appearing in large numbers in the archaeological remains of sites throughout the Ukrainian and Russian plains; by about 4500 B.P., they are found in western Europe and the Near East; by 4000 B.P., in Greece.

One can certainly argue that sheer numbers are not an automatic "good," but one cannot deny that reproductive success and extension of range are measures of evolutionary success. Today, wolves, panthers, and elephants vanish, ill suited, by their need for huge ranges, to the mere presence of man; dogs, cats, and horses flourish. From an evolutionary point of view, the survivors have done something right.

If the ancestors of domesticated animals had the biological motive to approach man, what of biological opportunity? What was the behavioral meeting-ground between man and wolf or man and sheep that allowed evolution to act on the biological motives of food and protection? This might seem a puzzle, because at least some of the behaviors that are the very basis of domestication, such as a lack of fear of novelty, were not present in the wild forebears of today's domesticates.

But it is a mistake to argue, as many long have, that domesticated animals behave in ways "contrary" to their natural instincts, a view that exaggerates man's importance in fashioning the behavior of domesticated animals. In herding sheep rather than eating them, for example, the sheepdog is often said to be working against its natural instincts. But in fact, in circling around the back of a flock of sheep—the essential trait of any useful herd dog—the dog is doing precisely what his wolfish ancestors did to cut off fleeing prey in the chase. Where a Border collie differs from a wolf is in not following through with the full sequence of hunting behaviors. It will run and chase animals and circle to cut them off; it will even, sometimes, grab a sheep by its hind leg.* But it will not go through with the kill, which is only the last step, after all, in a long sequence of behaviors. Its behavior is not opposite that of a wolf; it is merely incompletely developed.

The characteristics that are the sine qua non of domesticated animals, such as dependence and cooperativeness, are likewise not spontaneous developments; the raw material for these traits can be found in their wild progenitors' general way of life, one that can be called opportunistic. Dogs are scavengers; sheep and cattle graze a wide variety of environments; many domesticated plants have the basic characteristic of weeds in that they rapidly move in to exploit any cleared area. Biologists draw a useful distinction between two basic strategies for survival that an organism can adopt. "K competitors" live in environments such as tropical forests where competition is intense and every niche is filled to its carrying capacity. (In the mathematical equations describing populations, K is the symbol

---

* This results in disqualification at sheepdog trials, though is useful on the farm when one needs to separate out one animal. My Border collie will, on command, pounce on and hold a goose or hen, a talent that has eliminated many hours of frustrating chases.

that denotes carrying capacity.) Organisms in these environments survive by becoming highly specialized and milking every last bit out of their niche. "R competitors" by contrast live in relatively empty environments; they survive by adaptability and a high reproduction rate (denoted by the symbol R). Most domesticated animals are descendants of the latter, the dwellers of the forest edge rather than the forest, the scavenger or grazer that can eat a hundred different foods, not the panda exquisitely adapted to living off nothing but huge quantities of bamboo.

It is the edge dwellers, the opportunists, that would have been willing to frequent human campsites. They, not the conservatives, as zoologist Michael Robinson points out, are the pioneers. Robinson cites the interesting case of the blue tit *(Parus caeruleus)*, a bird that in its natural habitat is a forest-edge opportunist that feeds on a variety of insects and seeds. In the 1940s, 1950s, and 1960s in Europe, milk bottles that were delivered to the doorstep were topped with thin aluminum-foil caps. The blue tits quickly learned to apply in a new fashion the technique that they had evolved to search for insects in logs: They pecked holes in the caps and drank off the cream that rose to the top of unhomogenized milk. When my family spent a half year in Denmark in the early 1960s, standard equipment in all Danish houses was a cup that one would leave out for the milkman to place over the bottle top to keep the birds out.

The social structure of the wild progenitors of domesticated animals also in many cases provided a biological opportunity for domestication to take hold. Just as the Laplander and the reindeer share the social structure of nomadism that provides the meeting ground between the two species, so dogs, horses, sheep, cattle, and fowl all share a social structure with humans based on a social hierarchy.

All live in groups to begin with; thus living in a group with another social animal is not so great a jump as it would be for a solitary animal, such as a tiger. (Cats, which are solitary animals, are the one exception to this rule among domesticated animals; they are discussed in Chapter 5.)

All of the group-dwelling animals have a pecking order within their herds or flocks and have developed methods of expressing dominance and submission that are readily intelligible to us as well—and vice versa. Thus a dog understands perfectly well when we speak disapprovingly or slap it under the chin; we understand perfectly when it cringes, or growls. A horse, which outweighs a person by a factor of ten and could kill us if it so chose, can be trained to canter on command or stand at attention because of the natural instinct of subordination to a herd leader and because we share a system of communication that is mutually understandable. Although dominance hierarchies are sometimes initially established by fighting, the real biological point of it all is that it is a way to avoid the continual fighting that would otherwise occur for access to food or mates among animals that have to live together in groups. Once a dominant member has established his position, subordinate members will usually give way without a challenge. Thus a lower-ranked hen will allow itself to be pecked by a dominant hen without fighting back; a dominant horse need only lower its ears or make a menacing motion toward a lower-ranking animal to drive it away from a pile of hay. Submissive animals will approach a dominant animal in a manner designed to avoid provoking a fight. These behaviors are familiar to all dog owners: looking away, creeping with the tail between the legs, and, at the most extreme, rolling over and exposing the belly.

Because of these methods of signaling, dominance is

often established without any violence whatsoever. An animal that approaches another in a confident manner—or a mildly threatening one—will often be given way to as a matter of course. Skilled animal trainers have, through centuries of experience, developed methods that take advantage of this biological fact. Giving a command to a dog in a low, growling pitch, for example, is effective because it mimicks a dominance threat readily recognizable to the dog—the growl. Even when force is required, it usually doesn't take a human much to assert his dominance. Neither a choke chain on a strong dog nor a crop on a half-ton horse can physically force into obedience a willful animal bent on resisting our bidding; what they do instead is reinforce our position of dominance so the animal does not even try to resist. (Horses will often test new riders to see what they can get away with. Many hesitant or weak riders fail to establish a dominance relationship from the outset, through mistaken kindness, and wind up abusing their horses much more in the long run by having continual battles.)

Another type of group behavior found in sheep and cattle also may have provided a biological opportunity for association with humans: the propensity to follow a group leader into a new pasture. Studies have shown that the animal a flock or herd will usually follow is often not the dominant member of the group. For example, on dairy farms, one cow will invariably establish herself as the first to enter the milking parlor every day, and it is not always the dominant cow—though it usually is one fairly high in social ranking. Interesting for its implications for the domestication process, the leader will often be the one animals that acts independently—which could well be a human, if he has been accepted as a member of the flock. Nomadic herdsmen in fact often do seem to have taken on this role themselves; in Benjamin Hart's book, *The Behavior of Domestic Animals,* there is a remarkable photograph of a young herdsman in Africa briskly trotting ahead of a group of cattle being led to a new campsite. When starting out in the sheep farming business I once spent the better part of an hour in sheer frustration trying to chase a group of ewes into the barn; I'd get them moving, then one or two would break to the side and I'd run and try to cut them off. They got more and more skittish, I got more and more exhausted and disgusted, and I finally gave up and stomped up to the barn—only to discover upon looking back after a minute that close on my heels was a procession of the most docile sheep a farmer ever could desire. The biologist Valerius Geist, who lived in a log cabin in British Columbia to study Stone's sheep, notes that sheep will readily tolerate "harmless" humans in their midst. Geist reports that free-living bighorn sheep that became habituated to humans would follow him "at heel" for miles across open terrain.

The natural instinct for dogs, cattle, and horses to groom one another is another area where human and animal social structures overlap, and could well have aided the acceptance of humans as members of the herd or pack. Cows and horses will groom one another in areas inaccessible to themselves, such as the base of the tail. Dogs will clean out one another's ears. Hart observed nomadic Fulani herdsmen taking over this function; their cattle will approach the herdsmen and actually solicit his attention, and stand patiently while being scratched. Horses, dogs, and cats also clearly enjoy being petted and brushed by humans.

Upon all of these social meeting grounds the seeds of domestication were sown. But it would take the special climatic and evolutionary circumstances that prevailed at the end of the Ice Age to make them spring forth and flourish.

IV

THE SPECIES
THAT CAME IN
FROM THE COLD

Nature teaches beasts to know their friends.
—William Shakespeare, Antony and Cleopatra

If the behavioral makeup of man's would-be domesticates offered a social meeting-ground for the species, the peculiar climatic conditions of the world ten thousand years ago created the physical meeting-ground. Asking why domestication occurred exactly when it did, says David Rindos, is like asking why elephants appeared when they did; it is at a very fundamental level a question that is unanswerable, even meaningless. But there is nonetheless the tantalizing fact that domestication occurred almost simultaneously throughout the world. With the exception of the dog, whose appearance among the ranks of the domesticated preceded all other plants and animals by several thousand years, the first signs of domestication in the Near East, in Mesoamerica and Peru, and in the Far East all appear at about 9000 B.P., give or take a thousand years. Domesticated cereal grains appear at Tel Abu Hureyra at about 9500 B.P.; domesticated maize (though the cobs were tiny by today's standard, only an inch or two long) appear in the Tehuacán Valley in Mexico at least as long as 7000 B.P.; rice cultivation in China is thought to be equally old. The first domestic sheep and goats appear in the Near East within this same general period.

The discovery of this simultaneity once prompted all sorts of theories of diffusion. The people who "invented" agriculture presumably spread the good word. There was little evidence to support this idea, but it is an idea of undeniable appeal in part for the reasons elaborated earlier—it plays to our romantic image of the individual, heroic inventor. Thus the popularity, no matter how fanciful, of Thor Heyerdahl's notion of people sailing all over the world with blueprints for pyramids in their pockets to explain the simultaneous appearance of pyramids in Egypt and the Yucatán Peninsula. Although agricultural ideas, if not the people themselves, seem to have spread within some regions (such as from southwest Asia through Europe), there is no credible evidence for any connections among New World, Near Eastern, and Far Eastern agriculturists.

Debunking the diffusion theories does not necessarily mean that the simultaneous appearance of agriculture on distantly separated continents was all just coincidence, however. Something very unusual was happening both geologically and biologically at this time. The years around 8500 B.P. marked the end of an entire geologic epoch, known as the Pleistocene—a tumultuous million years in the history of North America, Europe, and Asia when glaciers repeatedly swept the continents and strange new animals came and went. The entire history of civilized man is compressed in the few thousand years that have passed since the end of the Pleistocene, and this small slice of recorded time has given us a very selective view of the earth and its history. The pine forests of Maine, the sweltering summers of the South, the rich farmland of the Midwest, are so much a part of our sense of the world that it is hard to believe they were not always so. In truth, they are an aberration from the geologic norm. Throughout the Pleistocene, a period that lasted more than a hundred times longer than our recent, brief era of civilization, the climate was in a state of continual and sometimes extremely rapid upheaval. As recently as eighteen thousand years ago, at the peak of the last major cycle of glacial advance, ice sheets covered about one third of the earth's present land area. What is now New York City lay beneath a mile of ice; glaciers reached as far south as Cincinnati, St. Louis, Kansas City, London, Amsterdam, and central Germany. In much of North America, Europe, and Asia expanses of open tundra stretched thousands of miles.

How might the climatic upheavals of the Pleistocene have set the stage for the rise of domestication? First, simply by forcing man and animals together. Across these treeless expanses, man and the wild horses, bison, mammoth, reindeer, and cattle he hunted lived on terms of intimacy that have scarcely been duplicated since. Archaeological evidence suggests that game were in an abundance hard to conceive of today, an abundance that must have seemed all the greater in relation to the human population. During the last major ice age, which began about thirty thousand years ago and ended with the final retreat of the glaciers to their current extent about eighty-five hundred years ago, the human population of Europe was almost insignificant. In *Ancient Europe*, Stuart Piggott cites the calculations of archaeologist Grahame Clark: Figuring the number of possible cave dwelling sites in England south of the line of permanent ice and drawing an analogy with modern hunting societies such as the Eskimos, who live in bands of about 25 to 50 each, Clark concluded that the entire population of the island might have been no more than 250—"a few busloads in a traffic jam." The population of what is now France, by similar reckoning, was probably no more than a few tens of thousand.

We are used to the idea of wild animals fleeing from us at the snap of a twig or at the whiff of human scent on a breeze, but that, too, would likely have been very different. All animals have an instinctive flight distance, the range that they will let a would-be predator approach before fleeing. It is to an animal's advantage to keep that distance as short as is minimally compatible with its continued existence; otherwise it will waste a great deal of time, and eventually wear itself out, fleeing from less than life-threatening situations. This is best demonstrated in the ironic fate that often befalls sheep worried by domestic dogs, notoriously inept hunters. If a sheep stood its ground against a dog, it might well survive. Instead, it usually runs until it drops dead from exhaustion.

Before the invention of firearms, the flight distance that wild animals would have maintained from humans must have been much shorter than it is today. The longer reach of firearms changed this: Animals that continued to maintain a flight distance more compatible with the bow and arrow era would not have survived to pass on this trait to the next generation.

Close contact between man and the animals with which he shared the compressed Ice Age territories south of the glaciers was a matter of the most ordinary routine. It is dangerous, if not impossible, to try to read the spiritual thoughts of men and women whose traces are a few paintings on the walls of a cave or a few carved pieces of antler bone. But there are some tantalizing hints of a religious significance with which Paleolithic hunters imbued the animals they knew so well and depended on so firmly for their survival. Whether or not the cave paintings represent a form of hunting magic, as was originally thought by their modern discoverers, they speak of an intimacy and acute understanding of the anatomy and behavior of the animals. Elsewhere, signs are clearer, such as the sites in northern Germany where reindeer had been sacrificed by tying them to stones and casting them into a pool. The later historical evidence from ancient Egypt that clearly domonstrates a worship of cattle is suggestive of more ancient antecedents of such beliefs.

There is something else that was biologically special about this time in laying the groundwork for domestication: The rigors of climate change in effect selected those very biological traits that would prove crucial to domestication. The million years that preceded the final retreat of the glaciers was not just a single Ice Age, but more correctly a series of ice ages, marked by enormous swings in climate that occurred, geologically speaking, with enormous rapidity. Glaciations occurred about every one hundred thousand years, alternating with interglacial periods when temperatures rose to within a few degrees of today's and the ice sheets retreated northward. Analysis of fossil pollen samples shows that the retreat of the glaciers was followed by the replacement of tundra and grassland plants with spruce forests and then oak. The ice sheets moved as much as two hundred feet per year at their swiftest. The entire process was accompanied by other environmental swings, in precipitation, ocean levels, and lake sizes. A series of smaller and faster fluctuations occurred within each major glacial period, but even these small cycles were on the order of several degrees over a few hundred years, much greater than anything in historical experience.

Raymond Coppinger advances the intriguing argument that such a traumatic fluctuation in habitat would have "penalized" any high degree of specialization within a species. As ranges expanded and contracted, as the forests moved northward behind the retreating ice cover and then retreated themselves behind the returning glaciers, a species that was locked into a single, highly specialized mode of existence would be less competitive. It is risky to make hard-and-fast correlations between extinctions and climate, but the entire glacial epoch was marked by the displacement of many animals from their accustomed ranges and by sudden extinctions that may have left many ecological niches unfilled. An enormous variety of new animals appeared—mastodons, mammoths, giant sheep and goats, saber-toothed tigers, huge-horned rhinos—and disappeared.

But some animals, needless to say, came through the Pleistocene with flying colors, apparently able to take quick advantage of expanding ranges in the wake of the retreating glaciers, adapt to a new environment, and fill these vacant niches. The wild sheep was one clear example of successful adaptation, spreading its range throughout most of the world during the intervals of glacial retreat. It has left modern traces of this march in what biologists term a cline: a more or less smooth geographical gradation in physical types from the point of presumed origin (the Barbary sheep of northern Africa) to their point of most distant advance (the bighorn of North America). As biologist Valerius Geist has shown, if one orders the dozens of different types of wild sheep found today across Asia and North America according to their physical similarities, the resulting line-up corresponds perfectly to their geographic distribution as well: Physical differences grow with distance. In other words, it seems likely that one form gave rise to the next, and that what we see today is the result of a dramatic spread in the range of wild sheep during the Pleistocene. What is more, species lines are hard to draw. Neighboring types are physically distinct, but the differences can be small; even the most distant, genetically and geographically, can interbreed and produce fertile offspring (even though some of the types have different chromosome numbers). All of which points to a fairly recent evolutionary origin for most of these species, and thus a fairly rapid advance in the wake of the retreating glaciers.

A species that had the ability to move rapidly into new territory and adapt to changing climate and vegetation was thus in a position to take advantage, in an evolutionary sense, of the fluctuating environmental conditions of the Ice Age. Adaptability is of course not a directly selectable genetic trait, like eye color. Rather, the ability of a species to adapt depends in part on the amount of natural variation that exists within the species. That variation, as Darwin demonstrated, is essential for the working of evolution: It ensures that a species is not locked into a single immutable form, and allows for the species to alter or new species to branch out when environmental conditions so dictate. A species of hare whose members come in widely varying shades from light to dark gray, for example, will have more genetic raw material to draw on if environmental changes—say, the advance of glaciers—suddenly make a pure white coat preferable for camouflage.

But the variation within a species is normally limited, both by natural selection itself and by basic rules of genetics and development. Over the short run, at least, the very fact that a species has an average size, for example, is a reflection of the fact that that particular size represents the best adaptation to current environmental demands. An individual within that species that is too far from the average will be maladapted. What is more, there are probably some very fundamental geometrical rules about the way genes are organized and the way cells divide that preclude extreme variations (such as individuals with three eyes). Any major departures from the prevailing average is likely to be a slow process; if environmental conditions favor giraffes with longer necks, for instance, it is not as if in one generation there might appear a couple of giraffes with 6-foot necks who will survive while the 3-footers perish. It is more likely that 3.001-footers in one generation will have an edge over the 2.999-footers; in the next generation the average may be 3.001, and a few individuals will be 3.002, and they in turn will be selected, and so on.

How can Darwinian evolution overcome this inherent speed limit on change? There is one source of enormous variation within a species that *can* be tapped—the change that all mammals and birds undergo in the course of developing from an infant to an adult. The range of variation that one sees within any adult population is miniscule compared with the differences that separate the average adult from the average juvenile. Both physical characteristics—everything from body shape and color to internal organs—and behavior change dramatically during development. And if adulthood is reached before the "normal" process of development is complete, some of the very strikingly youthful characteristics of the species will be locked in, while characteristically adult behaviors and structures are never developed or activated. Because the entire process of development is under genetic control, relatively small changes in the genes that determine the rate of development can produce enormous changes in the adult form.

This process is well documented in the physical evolution of many modern species, man included, and is termed *neoteny*. Human and chimpanzee babies, for instance, are in many respects strikingly similar in facial appearance, with high, domed foreheads and rounded cheeks; adult humans retain many more of these characteristics than do adult chimps.

The environmental changes that directly preceded the era of domestication would have been a powerful force favoring the selection of neotenic characteristics in many animals. First, neoteny is a way to introduce a whole slew of new traits very rapidly. It provides an abundance of raw material for specific, new traits that might be an advantage in a new environment; it is a way to overcome the inherent limits that natural selection itself imposes on the variation that is available within any species, in any one generation, for natural selection to act upon. But more important, the very characteristics of neotenates, even unrefined by any further selection, could themselves have proved advantageous in a changing world. All young mammals and birds show a curiosity about their surroundings, an ability to learn new things, a lack of fear of new situations, and even a nondiscriminating willingness to associate and play with members of other species—all of which are lost as they mature into adults and develop much more predictable and fixed sequences of behavior needed to hunt or forage, maintain their place in the social hierarchy, and compete for mates. Curiosity, a willingness to move into new territory, and an ability to learn throughout life are advantageous characteristics for individuals faced with a swiftly expanding or contracting ecosystem.

Studies of the evolution of wild sheep suggest that that is indeed precisely what happened. In both physical appearance and in behavior, those wild sheep today living farther from their point of geographic origin in the Pleistocene show progressively more pronounced neotenic traits. Geist observed that along the cline from the Barbary sheep, to the urali sheep of southwest Asia, to the argalis of central Asia, and then to the bighorns of North America, a number of juvenile traits become more pronounced. Horn size is the most noticeable. Larger horns in adult sheep are actually a "juvenile" characteristic: Growth is a juvenile trait, and the retention into adulthood of this trait results in adult rams that can continue to grow larger horns throughout their lives.

Geist argues that the pressures that select for large horn size in rams are a direct result of their moving into an expanding range. In the social structure of sheep, rams with small horns are automatically relegated to lower positions of dominance; they tend to assume a subordinate position toward larger-horned rams without fighting. Horns thus act as a signal or badge of social rank. Rams with equally large horns, however, have no way to settle dominance except by fighting. The net result is that rams with large horns are more likely to be victorious during the rut, and are likely to breed more ewes. While they're still alive, that is. Because the other side of the coin is that large-horned rams, because they are much more likely to be challenged and fight than are small-horned rams, are also much more likely to be killed in such fights. The longevity of the small-horned rams is a compensating factor. They may not breed as many ewes each year, but they are around for more years to breed.

In a stagnant population living in a habitat filled to capacity, these counterbalancing virtues are especially well rewarded. When sheep populations are high, food becomes scarce, pastures become filled with parasites, maturation rates are slowed, and the amount of energy any given animal has available to devote to fighting and reproduction is cut. As population growth slows, there will be few young, new large-horned rams to take the place of older, large-horned rams that are killed during rutting. Small-horned rams, by virtue of their longevity, may over time be just as successful in passing on their genes as their large-horned rivals.

But consider what happens to a population colonizing virgin territory. Food is abundant, growth rates are high, reproduction rates are high, and the burgeoning ranks of young large-horned rams will quickly provide replacements for any older large-horned rams killed in a fight. Mere longevity is no longer a virtue, when it comes to reproductive success.

Thus each time sheep spread into an uninhabited range, the pressures favoring larger horn size—and thus neoteny—would be repeated. As the migrating sheep filled up their new habitat, horn size would stabilize, but at a new, larger level. And these neotenized populations would then become the foundation stock for the next wave of dispersal, and further neotenization.

Wild sheep also show signs of progressive behavioral neoteny. The North American bighorn sheep *(Ovis canadensis)* act in ways distinctly more juvenile than do more primitive types they descended from, such as the Stone's sheep of British Columbia *(Ovis dalli stonei)*. Typically, as rams mature, their behavior becomes more "adult" in several distinct ways: They display their horns more frequently, kick subordinates with their front legs, and are less likely to mount ewes that are not in heat. In all of these characteristics, fully adult bighorn sheep behave in less adult ways than do Stone's sheep.

Coppinger suggests that the same argument could be made for other animals that expanded their range in the intervals between the glacial ages, including wolves, cattle, goats, camels, and rabbits. The very process of expansion would have favored neoteny.

This systematic neoteny, selected as a way to adapt to a changing world, would have laid an even more solid foundation for the interactions of humans and other animals. The curiosity, the lack of a highly species-specific sense of recognition, and the retention into adulthood of juvenile care-soliciting behavior (such as begging for food) of neotenates would all have been powerful factors in inducing wolves, sheep, cattle, horses, and many other occupants of the Asian and European grasslands of the late glacial era to approach human encampments and to allow humans to approach them. The neoteny that is part of our own evolutionary heritage may have likewise made us more willing to enter into relationships with animals other than the highly specialized one of predator to prey.

We have already seen, by looking at examples through-out the natural world, the advantages that can accrue to mixed flocks, and thus the biological motive that could have operated on animals already thrown together by bio-logical and climatic opportunity (compatible social struc-ture, neoteny, and physical proximity). What we have yet to do is explain precisely how the process of domestication could have been completed. That requires looking beyond motive and opportunity to biological *means*—the mechan-ics of evolution. An examination of another set of examples from nature can provide some valuable clues. For man is not the only animal to practice agriculture.

While it may seem a digression to examine the domes-tication of plants, the well-documented natural history of plants and the animals they depend upon to propagate themselves can provide evolutionary insight into the un-derlying mechanism of domestication of animals. In par-ticular, they can shed light on how a predator-prey relationship can evolve step-by-step into a much more equal symbiosis—and usually more as a result of adapta-tions made by the prey than by the predator.

In some of these plant-animal relationships, Rindos points out the "farmer" seems more an unwitting dupe than an intentional cultivator. The habits of squirrels and birds to cache nuts for the winter, for example, has pro-vided many trees with a means of propagating themselves. Studies of nutcrackers (*Nucifraga columbiana*), which feed upon pinyon pine nuts, show that in mild winters the birds may eat only a third of the nuts they store. Here is what might seem to be a predatory relationship, but it is clearly much more complex. Those trees that produce edible nuts that attract birds, and produce them in an abundance, would benefit from a mechanism that was sure to propagate its descendants farther and wider than trees whose seeds were not so attractive. The nutcracker is really just eating; it is not in any deliberate sense planting the next generation of trees to provide itself a food source. Still, the end result is a process that has many of the essential characteristics of farming.

A similar case in point is the tree *Calvaria major,* which produces a large edible seed with a hard coat that to achieve germination has to be preactivated by the grinding action of the dodo's gizzard. In this case, though, the tree seems to have bet on the wrong horse, or rather the wrong bird: Not one tree of the species has germinated since the dodo went extinct in the late seventeenth century, and now the tree itself seems doomed to extinction.

Ants are probably the most assiduous farmers in the natural world. The plants of at least 225 genera are solely dependent upon the activities of ants for their propagation. Ants will collect seeds and store them in underground chambers, affording a degree of protection that would not otherwise exist. Many plants have evolved special fatty appendages to their seeds that the ants will clip off to eat— after transporting the seeds to the vicinity of their nests to assure themselves a nearby supply of food in the next year's crop.

In their relationship with one species of acacia tree, however, ants take on a much more active part in cultivating and tending their crop. The acacias encourage the ants by producing leaf tips that are rich in protein and fat and which the ants clip off and use to feed their larvae; the trees also produce a sugary nectar that the ants feed upon. The ants live in special hollow, enlarged thorns. In turn, the ants assiduously remove the weeds in a wide circle around the tree, thus protecting their crop from competition and the brush fires that are common on the savannas where they grow; they also eat herbivorous insects and, as David Rindos reports from personal experience, inflict a painful sting on any animal that comes near. When the ants are experimentally removed, the trees die in less than a year.

As Rindos points out, there is no need to invoke any intent on the part of the ants to explain this. The ants don't literally protect the tree in return for food; they merely feed. The tree has evolved in an environment that includes ants and ants' behaviors; those behaviors include defending a home territory by stinging, eating weeds, and eating insects. A tree that offers something to attract ants, such as a totally dispensable leaf tip, will have an advantage.

One additional point is of the greatest significance in understanding the natural process of domestication. By entering into this symbiosis, the acacia can divert energy once expended on defense into growth and reproduction. Most species of acacia protect themselves by manufacturing poisonous chemicals known as alkaloids. Like man-made pesticides, they are complex organic chemicals that, from a biochemical view, are not at all cheap to manufacture. The acacias that are symbiotic with ants, however, do not produce any chemical defenses at all, which likely explains why they die so quickly when the ants are removed.

The connection between the evolutionary loss of defense mechanisms and the evolution of domestication is even more apparent in the symbiosis between aardvarks (*Orycteropus afer*) and a species of gourd that is regularly found growing at the entrance to the animals' tunnels. The gourd, *Cucumis humifructus,* is closely related to the muskmelon, and grows in desertlike regions of southern Africa. After pollination the stalk elongates and buries the ovary in the soil, where the fruit, which is large and juicy, matures.

The melon is the primary water source for aardvarks during the dry season, and saves the animals from making the dangerous trip to the waterhole.

The plant is exceedingly well adapted to the aardvark's habits. The seeds germinate poorly unless "pretreated" by passage through the gut of the aardvark. And since the aardvark buries its feces in much the same manner as does the domestic cat, the entire process ensures that seeds will be actively planted, and well fertilized.

The interesting point is that this melon is the only wild cucurbit (the family that includes many plants found in man's vegetable gardens, including cucumbers, eggplants, squash, and pumpkins) that lacks a natural, bitter toxin. Here is a case where the step-by-step evolutionary process of an agricultural relationship is quite clear. A melon that was slightly less bitter would be favored by foraging aardvarks; the seeds of such a melon, by virtue of their special care at the hands (or rather intestines) of the aardvark would be more likely to germinate and take their place in the next generation. Likewise, those aardvarks that fed upon the melons would be assured a safe and abundant source of water and would be more likely to survive, too. The process is purely mechanistic. The farmers in this case are, to be sure, favoring certain plants for consumption; they are not intentionally cultivating them.

Could the unconscious habits of the humans who were living ten thousand years ago similarly have led to the domestication of present-day cultivated plants? That is, is the relationship between humans and cultivated plants one of *biological* significance, dictated at least in part by the plant's adaptations to humans as a means of ensuring its propagation? There are some powerful analogies between the experiences of aardvark as farmer and man as farmer.

Under human domestication, many plants have lost natural defenses: eggplant, which is highly bitter in the wild form; lettuce, which has a high latex content and spines; cabbage, which contains poisonous glucosimates. The earliest domesticated plants were naturally invasive, and would have readily taken the lead in colonizing the areas surrounding human settlements. In the Near East, the sites where the earliest signs in the world of cultivated wheat and other grains appeared were at the very edge of the Mediterranean forests that had moved in during the postglacial period. A rich variety of wild cereals and legumes came right to the edge of settlements such as Tel Abu Hureyra; the remains of plants including wheat, wild rye, and various legumes have been found on the site, dating from the earliest days of settlement, around 11,500 B.P. Even today, in what is now Iran, Iraq, Syria, and Turkey, stands of the wild precursors of the first domestic grains (wild barley plus two kinds of wild wheat, emmer and einkorn) can be found growing. These wild stands were so dense that they could have been harvested quite easily—what one archaeologist has dubbed the lawn mower hypothesis of domestication. Combine that hypothesis with the so-called rubbish heap theory of plant domestication, and one has a striking echo of the aardvark-melon symbiosis: Plants that humans collected in the wild and scattered the remains of about a campsite would get a foot in the door; it would take relatively little human action to encourage their growth. The natural favoritism man would apply in gathering wild plants—for example, favoring those that had ripened simultaneously, those that had fewer natural defenses such as bitter toxins or thorns, or those having large seeds with thinner coats and seed heads less likely to shatter—would automatically steer the evolution of these plants in the direction of what we, in hindsight, term domestication, since many of the characteristics that make a plant worth gathering make it worth cultivating. The substitution of human cultivation for more "natural" methods of propagation would have, in the long run, proved a strategy for survival of evolutionary significance from the plant's point of view, just as acacias benefited by growing special features and abandoning the production of toxins in order to favor ants. The deliberate practice of saving seed and clearing land and sowing could have come much later—indeed after the essential steps in the process of domestication were complete.

The human behaviors of clearing land around a camp and throwing out the garbage, in other words, provided part of the environmental framework necessary for domestication of plants to proceed. Once more-deliberate cultural practices, such as saving seeds, began, an evolutionary context was provided for even more rapid domestication of other, new species. The power of plants to adapt to such human practices without any human intent whatsoever is clearly demonstrated by the very existence of such a thing as weeds. Weeds are a sort of parasite on the coevolved relationship between man and domesticated plants. For example, some weeds that have invaded flax fields *require* threshing to separate the seed and thus to reproduce. Needless to say, such a plant has absolutely zero survival potential in the wild. The action of clearing fields, harvesting, and threshing has provided the selective force necessary for this weed to evolve what, in absence of these actions by man, is a totally nonadaptive characteristic.

Teosinte is another interesting case. A relative of maize, it is indistinguishable from cultivated maize until the flowering stage is reached, by which point it has already gained the benefit of having been planted, weeded, fertilized, and tended, and when pulling it out would be far too much trouble. It, too, has characteristics, such as not dropping its seeds, that make it incapable of surviving in the absence of a human system of cultivation, harvesting, and seed-saving. Some later domesticated plants may well have originated as weeds of cultivation, a phenomenon sometimes termed "secondary" domestication.

Rindos adds the intriguing observation that most European weeds do very well in America, while native North American weeds do poorly in Europe—the point being that European weeds were adapted by thousands of years of coevolution to an "agroecology" introduced by Europeans into America only a few centuries ago. Many of the wild flowers, for example, that are now common throughout the fields and waysides of North America are weeds of cultivation introduced from Europe. They are well adapted to aggressively and rapidly colonizing disturbed ground; they flourish in the bright sunlight of open spaces. Native North American wild flowers, adapted to shady, moist woodland conditions, fared poorly by comparison as large expanses of land were cleared for European-style agriculture.

The argument that the relationship between humans and cultivated plants is a product of evolution is underscored by the fact that certain perfectly plausible plants have never been domesticated—at least by humans. Consider acorns: People have been using them as a food source for a long time. Indians on the west coast of America, along with inhabitants of Europe and the Near East, gathered acorns and prepared a sort of flour that, while it may not be very appealing to modern palates, made a significant contribution to the diet. But oaks, unlike many other nut trees, have never been domesticated. The answer that Rindos proposes to this curious riddle is that squirrels already had the job. The trees had already made the adaptations needed to encourage squirrels to propagate the species; it is hard to imagine any change in the acorn that would selectively discourage squirrels while appealing to humans. Nor, indeed, is it easy to see that a tree that had evolved such an efficient dispersal mechanism would gain any real advantage in having people planting its seeds instead.

The transformation of a predator-prey relationship to one in which both parties benefit has a fundamental effect on the demographic relationship between the predator and prey populations. In place of the boom-and-bust cycle of growing numbers of prey leading to growing numbers of predators, leading to a collapse in prey population, domestication allows both populations to expand. As Rindos explains in *The Origins of Agriculture:*

> Domestication permits larger numbers of prey to occupy the area by maximizing the probability that the prey species will be able to colonize any suitable open space in the environment. The prey oganisms best adapted to the behavior of the predator will prosper and in time will dominate the population.... the prey that maximizes the numbers of predators will be most fit.

Explaining the precise evolutionary steps that led to domestication of animals is harder. But one can imagine several ways in which the initial circumstances that threw man and animals together in this period, and the overlapping social structures of the species, could have been intensified by actions on both sides that were free of any methodical intention to achieve the result of domestication.

First and most obviously, there would be an automatic reinforcement of the neotenic traits of docility, curiosity, and disrespect of the species barrier. In exploiting the "open space" in the environment created by the refuse of human settlements and, later, by man's cultivated fields, an animal that is more adaptable and less fearful would be at an advantage. And since neoteny couples traits such as adaptability with docility, the selective pressures that man might apply to animals that encroached upon his place of dwelling—namely tolerating those that are tame, while driving off or killing those that are aggressive—would reinforce this selection of neotenic characteristics.

Some of these associations could lead very directly to symbioses. It is easy to see how cats, with their ability to control rodents in dwellings or granaries, would be encouraged. A less widespread practice, but which also points up how docility and a willingness to approach human habitations would yield payoffs on both sides, is found among the Itel 'mens of Kamchatka in northeastern Russia. They will rob caches of tubers collected by mice to attain a source of vegetables in the winter, but will encourage the mice's presence by replacing the stolen tubers with fish.

A very different source of reinforcement of neotenic traits, again without there necessarily being any conscious effort to produce domestication, may be found in the practices of hunters in the ancient Near East, probably including those at Tel Abu Hureyra, to drive entire herds of wild animals into corrals to be killed. A refinement over the methods practiced by North American Indians, who would drive bison or antelope herds over cliffs, the Near Eastern hunters built stone and boulder enclosures, some as much as several hundred feet across. An opening on one side was fed by two walls that stretched sometimes miles into the desert, fanning out in a wide V shape. The animals were driven into the opening of the V and stampeded into the corral, where they were killed. The structures, dubbed "desert kites" for their fancied resemblance to a toy kite when viewed from the air, are found throughout what is now western Jordan and north into Syria. The open end of the V often is situated along the valleys that the herds of gazelles followed in their annual migration. Gazelle bones from Abu Hureyra tend to confirm the suggestion that mass killing rather than selective hunting was practiced; the bones are of animals of all ages, and analysis of the teeth of newborns and yearlings indicate that they were all killed at the same time of the year.

If the practice was applied to some of the wild progenitors of domestic species, such as sheep, goats, or cattle, the corralling of large numbers of animals offers one mechanism for further selecting neotenic traits. Many more animals are killed than can possibly be needed at once. Though presumably much of the meat was salted and dried, corralling might well have raised the possibility of keeping some of the animals alive for future needs. Quite naturally, it would be the aggressive animals that would be killed right away, the tamer ones kept.

A final ingredient would have been needed to complete the biological process of change whose end result was the first stages of domestication: The incipient domestic population must have at some stage become isolated from its wild conspecifics. The classic problem for biologists seeking to explain how, according to Darwin's rules of natural selection, a new species branches off from another is that if the original species is well adapted to its environment, natural selection will tend to disfavor any individuals that start to differ to any significant degree from the original type. Individuals that march down from the "peak" of adaptiveness that the original species represents will, at least initially, be at a serious disadvantage. Even though there may be a second peak some genetic distance away, the problem is how to get there: You first have to march through a valley of relative nonadaptiveness. It is a paradox. Even when large genetic changes—changes so large that they constitute the formation of a new species—are advantageous, the small genetic changes that it takes to get there are, at least initially, distinctly disadvantageous.

Luckily, mathematics offers a way out of this dilemma. Even when natural selection is operating to keep the traits of a species, on average, at the peak of adaptiveness, variations of course occur. And in a small, isolated population, the simple laws of chance say that such variations, even when nonadaptive, are more likely to get at least a temporary foothold. Just as it is more likely to get all heads when you flip three pennies than when you flip one hundred, so chance events are more likely to cause the population to wander off its peak in a small population than in a large one. Selection is still a force, but it can be undone temporarily by chance. In subspecies that were geographically isolated from one another but which nonetheless share similar environmental conditions, small but distinct differences often appear for precisely this reason. The small founding population of an isolated subspecies will often reach a second adaptive peak despite environmental pressures that discourage wandering through the maladaptive valley.

Archaeological evidence does indeed suggest that it was in places outside the normal range of the wild progenitors that the first domesticates arose. Anthropologists, who tend to look for forces within human populations to explain the

adoption of agriculture, attribute this to growing population pressures that forced small bands to leave the optimal
areas where wild grains and wild game were abundant. In
an attempt to re-create the standard of living they were
accustomed to, they were thus forced to plant grains and
husband livestock. There is considerable appeal to this idea.
It explains the mystery of why humans would have deliberately adopted a way of life that was harder work and
provided a lower standard of living than hunting and gathering: Only those humans who would have been worse off
otherwise did so. Sites where some of the earliest traces of
domestic goats are found, such as Tepe Ali Kosh in southwestern Iran, were apparently just such "marginal" zones,
with a more arid and arduous climate.

The other way to look at it, from a strictly biological
point of view, is that regardless of the motivation of the
humans involved, it would *only* have been in areas at the
fringe of the natural range of the wild goat that the genetic
isolation necessary for a divergent branch of the species to
arise *could* have occurred. From the point of view of its
human inhabitants, the midwest of North America was
probably not a "fringe" or "marginal" zone about 4000 B.P.
when the first domestic sunflowers were cultivated there;
the inhabitants were not struggling to maintain a better
standard of living enjoyed by any more-optimally situated
neighbors. Yet the area is outside the plant's original range,
which was the western United States. Similarly, the tomato
and the chili pepper, which appear to have originated in
the Andes Mountains, first appear as domesticated crops in
Central America and Mexico.

But the central issue is that no matter what the intent of
the humans involved in the bargain, it would not have
taken very many members of a pioneering species—the

opportunists who were thrown together with humans by climatic circumstances and social predisposition—to establish the foundation of a new species, one that had exchanged elaborate defenses for dependence, adulthood for perpetual adolescence.

V

Y O U T H F U L
D E S I G N S

The youth, who daily farther from the east
Must travel, still is Nature's priest.
—WILLIAM WORDSWORTH,
"Intimations of Immortality"

Thirty years ago, a Soviet biologist, D. K. Belyaev, began an experiment in Siberia in the breeding of the silver fox, a dark-pigmented variant of the red fox *(Vulpes vulpes)*. The genetics of the fox, important to the Soviet fur industry, had been extensively studied; specific genes controlling coat color had been identified. But although silver foxes had been bred in captivity since 1892 on fur farms, and although some selective breeding for traits such as fertility and high fur quality had been practiced, the animals were not domesticated in any strict sense. They retained all of the essential characteristics of their wild counterparts. They molted and came into heat in a strict seasonal cycle, as in the wild; their behavior toward humans was no different than that of a wild fox raised in captivity.

So Belyaev decided to try an experiment. He would rigorously select animals for breeding applying but a single criterion: Those animals that showed consistently tame behavior toward humans would be kept; those that did not would be eliminated from the breeding program. Within five generations changes were already apparent. By 1979, twenty years into the experiment, the results were astonishing. His tame-selected foxes were not just tame; they acted for all the world like domestic dogs. They approached familiar persons and licked their hands and faces. They barked like dogs. They even sought the attention of strangers by whining and wagging their tails. Their annual molting cycle was disrupted, and the females began to come into heat twice a year, like dogs, and unlike both foxes and wolves. They also developed some physical characteristics of young foxes, such as drooping ears, and some of the variations in traits seen in other domesticated animals, such as piebald coat coloration.

Trying to deliberately breed for these characteristics one at a time—say, by selecting animals that showed slightly less erect ears—would have taken infinitely longer. And some of the characteristics of his domesticated foxes, such as twice-a-year estrus, would seem to be almost impossible to generate through such a conventional approach to selective breeding. They are entirely new traits that simply did not exist in the wild foundation stock.

In the classic view of Darwinian evolution, and in conventional notions of artificial breeding as practiced by man, the only source of new traits is the natural variation within a species (e.g., some animals have floppier ears) or the quite rare appearance of a chance mutation. In fact, to produce any radical change through a conventional breeding program, mutations are almost essential. The breed of sheep known as the polled Dorset, for example, are the descendants of a single hornless mutant that appeared in a flock of horned Dorsets. Breeders might have tried painstakingly selecting in each generation the members of a flock that had the smallest horns, but the process would have been agonizingly slow, and even then of dubious outcome.

Belyaev in an astonishingly short time produced not just one new trait, but a whole package of new characteristics. What he had done, by selecting for nothing more than tameness, was to tap into the same powerful evolutionary tool that nature had employed as a solution to the successive ecological catastrophes of the ice ages: neoteny. No sophisticated understanding of genetics was required. No elaborate breeding schemes of crossing and inbreeding and hybridizing were involved. It is a feat that the first agriculturists, carrying out a plan no more profound than driving away or killing off aggressive members of a herd while tolerating the tame ones, could easily have achieved. What nature began as an adaptation to the Pleistocene, man completed, however unconscious he likely was of the ultimate outcome. Neoteny was the tie that bound evolution to domestication.

The power of neoteny as the guiding force behind domestication is evident even today. Virtually all of the important characteristics that set apart domesticated animals from their wild progenitors can be accounted for by neoteny. For all of the sometimes fanciful selective breeding that man has practiced in the interim to satisfy his tastes for swift racehorses, fine-wooled sheep, or golden-coated dogs, the basic physical blueprint of domesticated animals remains faithful to an evolutionary plan drawn thousands of years ago. In physical appearance, especially in facial structure, nearly all domesticates resemble the adolescents or infants of their free-living ancestors more than they do the adults. They have shorter muzzles and high-domed, rounded heads. In some highly bred dogs, such as the Pekingese, physical neoteny has been carried to the point that they retain into adulthood characteristics actually found in *fetal* wolves: big eyes, extremely short-ened faces and large heads, curly tails, and soft fur. But even the very earliest domesticated dogs show the begin-nings of this process in their shortened jawbones and com-pacted teeth.

Horses and cats are the two domesticated animals that differ the least in physical appearance from their wild rel-atives; they are also the most recently domesticated. Even horses, though, may have some neotenic physical traits, such as their relatively long legs and necks that give them an overall proportion that more closely resembles a wild foal than a wild horse.

But it is the permanently juvenile *behavior* of domesti-cated animals that most clearly distinguishes them from their wild relatives, and which is the clearest testimony to the role of neoteny in their evolution. When Belyaev's foxes lick the face and hands of a familiar person, they are not just resembling dogs: They are also, and more to the point, resembling a fox cub or a wolf pup, which employ such "care-soliciting" behaviors to beg for food from an adult. A young wolf pup will vigorously lick an adult in the corners of the mouth to get him to regurgitate food.

That does not mean your pet dog is expecting you to vomit when he gives you a big slurp on the face. But he is acting out a juvenile behavior expressive of dependence and a need for care and attention.

Domestic dogs retain many other juvenile behaviors. Submissiveness toward other members of a pack of course occurs even in subordinate adult wolves, but the extreme submissiveness of domestic dogs toward their owners and their sometimes excessive demands for attention are often more characteristic of what is seen in pups. Rolling over and exposing the belly, whining, and averting the eyes are all readily recognizable juvenile behaviors that dogs retain throughout adulthood. Many dogs like to paw at their owners to get attention. My collie will even raise a paw in the air when he sees me coming. This is another juvenile behavior, which may have its origins in the action puppies use on their mothers' udders to stimulate them to let down milk. It also is part of the ritual that a submissive dog will sometimes use when approaching another—it will lift up one paw and place it over the neck of the dominant one. The retention of this juvenile or at least submissive behavior in dogs explains, incidentally, why it is so easy teach a dog to "shake hands."

Behavioral neoteny also helps to explain why cats, which, unlike all other domesticated animals, did not live in groups in the wild, are willing to put up with human company. The one point in their lives when cats are social is in infancy and adolescence, when they live with their mother and littermates. (Lions are an exception; most cats, including the European and African wild cats *(Felis silvestris silvestris* and *Felis silvestris libyca)* that are believed to be the ancestors of the house cat, are solitary as adults.) A domestic cat pacing back and forth with its tail held straight up in the air while its owner opens a can of cat food is duplicating the gesture of a kitten begging its mother for food. The adult domestic cat's willingness to play and be groomed and petted are traits that are found in wild kittens but which disappear in wild adults. Another juvenile behavior that is retained into adulthood in domestic cats is the immobility reflex. When a mother cat needs to move her kittens, she lifts them up by the scruff of the neck with her mouth; the kittens respond instinctively by going limp and raising their hind legs and tail. By not struggling and by obligingly lifting up appendages that would otherwise drag across the ground, the kitten's response makes the mother's job that much easier. But adult domestic cats respond the same way, a useful fact that is sometimes employed by veterinarians who need to administer a shot or insert a rectal thermometer. It is unlikely that an adult bobcat would respond similarly to someone who seized it by the back of the neck.

Less well defined but still unmistakable care-soliciting behavior can be seen in horses, cattle, and sheep, which will often nudge a person with their noses at feeding time. My sheep will often let out a bleat that resembles a lamb's call for its mother when I open the barn door on my way to the feed room.

Finally, the propensity for all domesticated animals to treat humans as members of their own species, as opposed to either predators or prey, is a reflection of their systematic juvenilization. The highly species-specific behavior of a wild adult is an essential ingredient not only in hunting and protecting one's young but in rituals of social conduct such as territoriality and courtship. Young mammals, however, tend to be far less discriminating. They will play uncritically with members of another species if offered the chance. It is commonplace to speak of a dog that thinks it is human, but it is much more likely that the dog thinks humans are dogs. Horses, sheep, or cattle that are approached nonchalantly in a field by a person generally react with the relative indifference that they accord members of their own species, whereas they often treat an approaching dog as a predator, and respond by fleeing (in the case of sheep) or by becoming aggressive (in the case of horses).

Many of these traits that so conveniently come packaged together via neoteny are essential requirements for successful domestication. A farm or house animal has to be willing to put up with being touched and handled by humans. It cannot manifest behaviors such as aggressiveness, territoriality, and strong parental protectiveness if it is to live in peace in a farmyard. It cannot be excessively fearful of new situations.

As we have seen, the need to adapt to a rapidly changing environment created a natural selective pressure that favored neoteny as the Pleistocene progressed. In the early stages of domestication, when animals were still free-living and when humans would have been exerting their own selective pressures only loosely, by driving off or killing aggressive animals while tolerating tame ones, "natural" selective forces might also have favored neoteny in another way. Geist's explanation of neoteny in wild sheep during the ice ages is based on the fact that in a rapidly expanding population that is colonizing new territory, a premium is placed on sexual competition. In the case of sheep, that means the development of large horns, which in turn means neoteny, since "juvenilized" sheep retain into adulthood the propensity for rapid growth of their horns.

A similar process could have likely taken place in dogs or other animals that entered the new niche created by living with humans. The situation, ecologically speaking, was identical to that of sheep that move into a virgin pasture. The physical and behavioral niche of domestication was one in which food was abundant, competitors few. An immediate evolutionary advantage would have been conferred to those individuals that could reproduce quickly. While in a stagnant population there may be an advantage to longevity and slow development, in a booming population the advantage goes to whoever can fill up the niche with his offspring the fastest. That leads to a powerful selective force in favor of animals that reach sexual maturity earlier—which is the very definition of a neotenic animal, one that reaches sexual maturity before the full development of ancestral adult behaviors can be developed and activated. The young age at which dogs become sexually mature compared with wolves, and the fact that female dogs come into heat twice a year compared with once a year for wolves, is modern testimony to this bit of evolutionary history.

For exactly the same reasons that nature exploited neoteny as an evolutionary tool in the Pleistocene, humans apparently exploited it once they began more conscious selective breeding in the later stages of domestication in the Neolithic and beyond. Neoteny offered a rich source of variation that could then be drawn upon to produce breeds tailored for particular purposes. The huge variation that is seen from breed to breed within most domestic species (chihuahuas to St. Bernards, ponies to draft horses, wool sheep to mutton sheep) points to a breeding program that operated far more swiftly than one based on laborious selection trait by trait.

In particular, detailed studies of dog behavior and genetics suggest that the great differences between modern-day dog breeds can be accounted for largely in terms of the degree to which they have each been neotenized. Compared with the fully developed wild type—the adult wolf—all domesticated dogs are juvenilized. That was the first step to domestication, one we may roughly call nature's doing. But some dogs are more juvenile than others, and that was the second step, which was man's doing.

To Raymond Coppinger, this became apparent when he began a project in 1976 to explore alternatives to predator control, especially for the American sheep industry, which loses half a million sheep a year, mainly to coyotes and marauding domestic dogs. Coppinger set out to explore the possibility of introducing livestock-guarding dogs, which have been widely used by European shepherds for centuries. Despite the large geographical differences in their places of origin, and despite differences in details of appearance such as coat color and tail carriage, all the Old World guarding dogs that Coppinger encountered in his travels bore striking similarities in both appearance and behavior. They are all large, weighing about one hundred pounds when full-grown. Yet they all literally look like puppies, with floppy ears, large, rounded heads, and short muzzles. They are generally placid and unresponsive, but are "attentive" to the sheep they guard, moving with them and almost at times seeming to blend in with the flock. They appear to prefer the company of sheep to anything else. "As long as the routine was unbroken," Coppinger wrote in 1982 of his first encounter with the Shar Planinetz dogs in southwest Yugoslavia, "the dogs remained totally wrapped up in their own world. When, however, one intrepid biologist decided to see if he could 'steal' a sheep, and slowly but steadily approached the flock, he found his way slowly but steadily barred by one of the dogs." The local farmers said they did have wolves, but that the dogs could easily fend them off.

Returning with a good selection of pups to Amherst, Massachusetts, where he directs the Livestock Dog Project at Hampshire College's Farm Center, Coppinger began a breeding program to provide dogs to farmers in the United States for a small lease fee. Watching the pups as they developed gave a clear insight into what, biologically speaking, accounted for their extraordinary behavior. Simply, they are cases of arrested development. As Coppinger explains, once past the infant stage, when all pups are blind, deaf, and totally dependent upon their mothers, pups will enter a period of adolescence that lasts one to two years. In this period, they begin to develop a series of instinctive motor patterns that when carried completely to maturity in the dog's wild ancestors (and which can be seen in wolves today), results in a full-fledged hunting ability:

> The first stage is characterized by sitting outside the den and loosening the tactile contact between littermates and mother. Food is mainly obtained by licking the mother's face, getting her to regurgitate, and some fighting over the spoils. Pups are wary of novelty and the slightest stimulus will cause them to scurry for the den, yelping, or in our observations sometimes to crouch, snarl and bite in fear.
>
> In the second stage, pups begin to play with objects. At first the object may be a sibling or a parental tail, then later, a stick, a leaf or a bug. As this stage continues, the pups spend a greater amount of time playing with various objects.
>
> In the third stage, pups add a new motor pattern, that

of stalking. They lie in wait and pounce on siblings or even motionless objects. Sometimes the object is imaginary, or a pup will scratch a leaf or stick into motion and then pounce on it with a stiff-legged jump similar to that observed in coyotes and wolves. Later during this stage, motion can stimulate this behavior; a short chase can ensue to cut off the motion, a behavior known as heading.

In the fourth stage, pups start following a parent and may even participate in a hunt. During the hunt they begin to focus on the heels, if the prey is large, a behavior known as heeling. When all the behaviors from the four stages are put together with learning, the wild type can hunt successfully.

What Coppinger observed with his guarding dog pups is that as they matured they never really advanced past the first stage, marked by association with littermates, a suspicion of novelty, and fear-biting when provoked. When raised in the company of sheep, they readily transferred this behavior from their littermates to the sheep. Coppinger demonstrated the point by contrasting the behavior of a guarding dog in the kennels at Hampshire College with the behavior of a Border collie, a herding dog that is much closer to the adult wild type in its development of hunting behavior. Waving a ball back and forth in front of the guarding dog and encouraging him with his voice evoked no interest whatsoever in the ball; the dog did not even follow it with his eyes. Instead he became visibly more and more tense and defensive, backing away while raising his hackles at the same time. The Border collie, by contrast, keenly followed every motion of the ball and ran furiously to chase it. When Coppinger substituted a piece of gravel, threw it, and then kept egging the collie to go get it even after it had come to stop, the dog resorted to pawing the ground until she kicked up a piece of gravel and then whirled rapidly around to chase after it.

While guarding dogs are so highly neotenized that they are arrested in the first stage of adolescent behavioral development, Border collies are probably in the third stage. They will stalk and circle to head off retreating animals, but they usually do not show interest in following through with the kill. Dogs that are arrested in the second stage, namely object play, include retrievers and poodles. They will often exhibit an obsession with objects similar to the obsession that Border collies have with chasing. A few dog breeds are somewhere in the fourth stage of development, closest to the wild type. Welsh corgis, which are used to drive livestock from behind, show definite heeling behavior in nipping at the legs of their retreating charges.*

In one sense it seems an amazing coincidence that evolution provided so handy a means as neoteny for producing so many of the essential characteristics for the domestication of animals in one fell swoop. But that is backward thinking: There is no reason to believe that domestication was inevitable. Knowing what we now know, we can with hindsight

---

* Mutts do not fall into any one of these categories, but often share behavioral characteristics with guarding dogs. Coppinger suggests that by recombining the various innate motor patterns of their parents, mutts end up with a rearranged or disrupted sequence of predatory motor patterns. The Navajo shepherds of the American Southwest, who learned about the use of livestock-guarding dogs from the Spanish, have for centuries used mongrels in this capacity with good results. It is also interesting that mutts, like many of the first-stage dogs, make good pets. They form a close bond with their owners and are happy to hang around the house all day.

marvel that evolution and genetics should have rigged things so that the traits that make for a successful domesticated animal could have been produced so swiftly merely by selection for a single characteristic, tameness. On the other hand, were it not for the evolutionary mechanism provided by neoteny, it seems extremely unlikely that any animals would ever have been domesticated, and no one would be sitting around today scratching their heads and wondering why not. Domestication seems natural only because it happened, but it happened only because it was natural.

The significance of neoteny in the evolution of domesticated animals has sometimes been confused with the fact that humans find neotenic traits attractive. In focusing on this fact, interesting as it is, some researchers have assumed that neoteny in domesticates arose specifically from humans' having deliberately favored young-looking or young-acting animals in the course of selective breeding.

Some quite serious authors—and by no means just animal rightists—have gone so far as to suggest that our selection for neotenic traits is part of some darker human urge to trivialize animals. Elizabeth Lawrence, a veterinarian who has written extensively on the relationship between humans and animals, sees a connection between our breeding of neotenic traits into dogs and our popular cartoon images of animals as cute and cuddly:

> Among the explanations for why we in Western culture, and particularly contemporary American society, neotenize our animals as we do is our need to *gain a sense of control* over them. As docile and playful "children," they may be relegated to a separate category, without full citizenship in the world.

Lawrence even sees a dark message in a cartoon featuring a "serious and adult-proportioned" father bear phoning the insurance company to see if his policy covers stolen porridge and broken chairs:

> Humor in this case arises from the sophisticated response of the bear. Ironically, the childhood petulance of whining about loss of food and intrusion has been replaced by a cool, calculated adult response to theft and trespass. We find this funny precisely because according to our way of viewing the world, it is not going to happen.

Well, neoteny or no neoteny, few bears are ever likely to phone insurance companies, but Lawrence's arguments still are worth considering, for they raise some important questions about when in the process of domestication human intentions became important. A closer look at the psychological and behavioral evidence suggests that rather than a matter of conscious intent or a desire to gain a sense of control, though, the very real human preference for juvenile-looking animals is innate. Thus in the early stages of domestication, unconscious human preferences could well have reinforced the selection of neotenic traits that originally arose as an adaptation to a rapidly changing environment.

The behaviorist Konrad Lorenz was the first to suggest that so-called innate releasing mechanisms—stimuli that automatically elicit certain behavioral responses—operate in humans just as they do in many other animals. There is plenty of anecdotal evidence suggesting that certain physical characteristics of the human infant release an innate parental instinct. These include a short face and a large forehead, large and low-set eyes, protruding cheeks, short and thick limbs, and uncoordinated movements. Lorenz, to buttress his case that there is something innate about our affectionate and nurturing response to such traits, pointed out that even when they show up in very unlikely places—even in nonhumans—they elicit a similar response. We undeniably find animals with big rounded heads and short faces "cute."

Biologist and essayist Stephen Jay Gould has shown how the principle was taken to heart by the Walt Disney Studios. Mickey Mouse, in his original incarnation, was a rather mischievous, even cruel, character. But after the decision was made to turn him into something much more lovable, his physical appearance underwent an unmistakable juvenilization. Gould, with the aid of a pair of calipers, actually measured and plotted the progressive neotenization of Mickey over the years. In three successive stages, Mickey's eye size grew from 27 to 42 percent of his head length; his cranial vault and overall head size grew as well. Overall, he grew more and more like his young nephew Morty. The villains, on the other hand, inevitably are much more adult in appearance, with long snouts and small heads relative to body size. To quote from Gould's analysis: "In 1936 . . . Disney made a short entitled *Mickey's Rival*. Mortimer, a dandy in a yellow sports car, intrudes upon Mickey and Minnie's quiet country picnic. The thoroughly disreputable Mortimer has a head only twenty-nine percent of body length, to Mickey's forty-five, and a snout eighty percent of head length, compared with Mickey's forty-nine."

Doll makers and stuffed-animal makers have long applied the principle, too, churning out products with exaggerated juvenile characteristics. There is also anecdotal evidence indicating that people who have pets as child substitutes tend to select breeds that have more extreme neotenic traits.

Thus it is probably quite true that as animals began to acquire the package of neotenic traits that made them suitable for domestication, they also acquired a physical form that tended to elicit favorable, care-providing responses from humans. That is a long way from saying that humans selected animals solely on the basis of neoteny for neoteny's sake, however, or, even less likely, that it was all the part of some grand attempt to trivialize animals.

In the same vein, some biologists have argued that domesticated animals are "degenerate" forms, debilitated through selective breeding that they are no longer adapted to life in the wild. Unquestionably, domesticated animals are by their nature permanent juveniles, dependent upon us for care, for the protection from predators that they can no longer provide themselves, for the food that they no longer know how to find for themselves. But it can be argued just as well that they are supremely well adapted forms, adapted to a changing environment in a way that highly specialized wild animals are not. As much as we may cherish the image of majestic animals such as leopards and giraffes making their way in the jungles and the plains, their habitat is rapidly vanishing, in large part as a result of the very evolutionary success of man and the animals that have, through domestication, become dependent upon him. Domesticates have honed the traits needed to survive in a world that includes man and the changes his biological success has wrought. Their solution to the problem of survival is testimony to the remarkable resources of the evolutionary process; it is at the same time a humbling testimony to the less than complete control we exert over the world, our myths about ourselves notwithstanding.

VI

NO TURNING
BACK

*I discovered, though, that once having
given a pig an enema there is no turning back.*
—E. B. WHITE, "DEATH OF A PIG"

I f the rise of agriculture was not an ideological revolution, it was perhaps a slow subversion that, once
begun, could not be stopped. Human society had changed
in demographics and structure; the animals that entered
into an alliance of mutual dependence with that human
society had changed in biology and behavior; the environment, too, had changed—first by forces purely natural as
the glaciers retreated and the grassland habitat of many of
the domesticates' forebears vanished before the forward

march of oak forests across the landscape, then by forces artificial as a society built upon tillage and pasture slashed, burned, and uprooted those forests to meet its growing needs. If there was nothing inevitable about the emergence of agriculture, there was everything inevitable about its spread, its triumph over older ways of subsistence, and its irreversibility.

The triumph of agriculture was part of a larger story of evolutionary change, in which nature itself was redefined. The template for evolution for at least the last ten thousand years had included man and the environment created by his agricultural symbioses. The romantic conception of nature that is still with us, one in which noble, independent creatures proudly live free in unspoiled splendor, is almost a sort of racial memory of our hunter-gatherer past. The nature of nature was forever changed by agriculture, even if our conception of nature was not. The evolutionary reality for animals and plants now is a reality that, for good or bad, includes man.

As evidence from archaeological sites throughout the Near East makes clear, the initial adoption of agriculture was slow. At Tel Abu Hureyra, permanent settlements appeared at least a thousand years before the first grains were sown; another thousand years passed before sheep and goat herding took the place of hunting. At sites such as Jericho on the West Bank and Beidha in Jordan, hunting persisted for a thousand years or more alongside animal agriculture. In Europe, increasingly specialized hunters and fishermen flourished for millennia alongside the first farmers, who by all evidence worked harder, ate worse, and suffered more. Their bones bear scars of malnutrition, disease, and brutal burdens. The relatively carefree existence of the hunter-gatherer, who had his pick of scores of foods (the Kung, who carry on the hunter-gatherer tradition hunt more than sixty different species of animals, ranging from hare to buffalo, and know more than a hundred species of edible plants) and who (again, if the Kung are a reliable guide) spent no more than a dozen or two hours a week working to supply his needs, was replaced by an arduous existence of clearing land, hoeing fields, tending stock.

But agriculture offered one advantage: It produced more food. It was food ill suited in many ways to the nutritional needs of primates who for at least two million years lived as hunter-gatherers, but there was more of it—and more per square mile, too. Bands of hunter-gatherers need something like ten square miles per person; farming communities average a tenth of a square mile per person. And that intensity of settlement and the growing population that intensive food production soon sustained guaranteed that the old way of life was doomed.

One reason was simply sheer force of numbers. As competition for land intensified in the few millennia following the rise of agriculture, the more populous farming communities would have had the advantage in the ensuing warfare. In central Europe, fortified settlements surrounded by ditches and heavy pallisades built of tree trunks appear by about 5500 B.P.

Another was a wholesale transformation of the environment that doomed the previous way of life. It is fashionable these days to decry the despoilation of the environment by man as if it were something that had happened only in the last hundred years—and as if the solution were as simple as undoing the excesses of the Industrial Revolution. But nothing man has done has so drastically altered the face of the planet as the wholesale clearing of land for agriculture that accelerated around five thousand years ago. Hunter-gatherers who follow game and who roam across the land searching for food have a fundamentally limited impact on the environment. The effect is spread across a wide area from season to season; and the population itself is limited by available resources. But once a permanent settlement becomes possible, the pressures humans can bring to bear on the environment are focused and concentrated. Having made the commitment to year-round habitation in a fixed spot, the first agricultural peoples of the Near East were able to make an investment in permanent facilities that made their hunting much more efficient and deadly. Their desert kites, the stone-walled pens into which they drove herds of wild gazelles, allowed for mass killings on a scale beyond the reach of any wandering tribe of hunters. In the period from 9000 B.P. to 8000 B.P., kites appeared across the steppes of what is now northern Jordan; entire chains of them, extending tens of miles long, were built along the gazelle's migration routes through the valleys. The intense pressure thereby placed on the herds may, as archaeologists Anthony Legge and Peter Rowley-Conwy suggest, have finally forced the transition at Tel Abu Hureyra to sheep and goat herding around 8500 B.P. It would also have spelled the end of the way of life for any neighbors who had tried to maintain a pure hunter-gatherer existence.

The adoption and spread of agriculture forever altered the environment in other ways that made it impossible to turn back the clock. Not only was the game that had sustained hunters killed off directly; it was also being forced out indirectly by the clearing of forests in Europe and Asia to make way for planting and grazing. That began around 6500 B.P. in Europe, when the spread of agriculture and the demand for new land exhausted the available supply of lightly forested land near rivers, and a more aggressive assault on the environment was launched. Analysis of preserved bits of pollen shows that agricultural settlements established during this period in northern Germany, the southern Netherlands, and Czechoslovakia were on heavily wooded terrain. Only in areas such as Hungary and the Ukraine did open grassland remain from the retreating glaciers.

Experiments in ancient archaeological methods carried out in the 1950s in Denmark, combined with fossil pollen analysis, clearly point to the widespread use by these settlers of slash and burn (the same technique that, practiced today in South America, earns universal condemnation for its environmental unsoundness) to strip the land of its forest cover. In the Danish experiment, which attempted to reconstruct the practices of these first agriculturists, a patch of oak forest was cleared using flint axes. Half of the plot was burned, the other half not. In the unburned area, bracken, sedge, and grasses—almost exactly the same sort of undergrowth that had occupied the area before the experiment—gradually reinvaded the plot. But the burnt area was quickly colonized by very different plants—what we today would immediately recognize as weeds of cultivation: daisies, plantain, dandelion, thistle. These weeds show up throughout the fossil pollen record of northwest Europe at precisely this time, increasing in their abundance in lockstep with the decline in abundance of forest-tree pollen.

By about 4500 B.P., competition for land had apparently driven agriculturists to the higher plateau land, with its still denser oak forests and heavier soils, and the environmental transformation accelerated. The first plows begin to appear in the archaeological record at about this time, as do the earliest remains of plowed furrows in excavated soils. It is no exaggeration that the landscape of Europe and Asia has never been the same since. Agriculture sowed the seeds that ensured the destruction of any competing systems. It did not even matter if agriculture was a superior "invention"; by its very nature, it wrought changes in the surrounding environment and in the human social structure that guaranteed the elimination of other ways of life. The near extermination of the American bison by white hunters in the nineteenth century tends to obscure the fact that it was the arrival of European agriculture in the New World and the resulting appropriation of the bison's habitat for grazing and the destruction of its habitat for tillage that really sealed the bison's fate—and the fate of the Plains Indians who depended upon it. Even after what is universally considered a superbly successful conservation effort in the last century, the bison survives today on the merest fraction of its original range.

Finally, the same environmental changes that had provided the selective pressures favoring coevolution toward domestication—such as the disappearance of the tundra habitat that had once sustained vast herds of free-living horses—had in no way diminished with the rise of agriculture. The animals that had for so long sustained the hunting bands were vanishing, victims of a purely natural evolutionary force beyond their control or man's control.

Another force guaranteed the irreversibility of the mutual dependence that grew between humans and domesticated plants and animals, and it is the ultimate comeuppance to the traditional view of agriculture as a milestone in human achievement. Anthropologists have always assumed that agriculture was a way to ensure food supplies, overcome the vagaries of nature, and bring production under human control. But David Rindos suggests that it was the very *instability* of the food supply from agriculture that secured its foothold. "Agriculture," Rindos says, "is a disease spread by its own symptoms"—like a cold virus spread by its victim's sneezes.

Although a hunter-gatherer society can shift from one food source to another when drought or flood, heat wave or frost hits, farmers throughout history have tended to rely on a handful of crops, or more often a single, staple crop. That is true even today: Such staples include maize in North America, wheat or other cereal grains in Europe, rice in Asia. The inevitable fluctuations in weather take a regular toll on any agricultural society. Climatologists are fond of noting, when assailed by farmers (or picnickers) demanding to know what is going wrong with the weather, that every day, somewhere in the world, a record is being set. Average temperature and precipitation is not the norm; deviations from the average are the norm. Even in this day of irrigation, hybrid seed, fertilizers, pesticides, and machinery, nothing so influences a farmer's yield as the vagaries of the weather. The history of agriculture has been one of good and bad luck, of boom and bust, of golden ages of prosperity (such as the halcyon days of the 1910s in America, a period of unequaled income for farmers) and dark ages of crisis and famine (such as the 1920s and 1930s, when the economic instabilities of the Depression were multiplied many times over in the agricultural sector by devastating drought).

The effect of such fluctuations strikes especially hard in a community that has enjoyed prosperity, and whose population has been rising. In general, a population will rise to keep up with the growing carrying capacity of the environment. But a bad year can suddenly leave a rising population stranded above the carrying capacity. As Rindos points out, a people used to abundance will see a sudden drop in production as a crisis even if it stops short of outright famine. And in those circumstances the natural solution is for a portion of the population to emigrate in search of new land—where they will reproduce the same way of life that got them in trouble in the first place. Such a pattern has been repeated as recently as the 1930s in the Dust Bowl, where thousands of farmers from the Midwest moved westward in search of new opportunity; it is not difficult to imagine a pioneering Neolithic band similarly striking out on their own from the famine-stricken settlement of their birth. Agriculture would have created rising populations with rising expectations that would repeatedly and inevitably have been dashed, and which would repeatedly and inevitably have created an appetite for ever more land. Once in this cycle, man would have found it hard to break.

There was no breaking it for the plants and animals that had driven the entire process, either. The fundamental genetic changes that had transformed free-living, highly specialized organisms into dependent and despecialized members of a complex symbiosis erected an equally formidable barrier against evolutionary revanchism. Many domesticated plants are totally incapable of surviving in the wild. The seeds of maize, flax, and to a lesser degree the grains are released only by mechanical threshing. In the wild, they would be incapable of reseeding themselves efficiently. And because they have been bred so the entire crop germinates simultaneously once the seeds are sown, domesticated plants in the wild can be rendered extinct by one ill-timed drought. They are as dependent upon us as we are upon them.

While this may be less true for domesticated animals, some of which can scratch out a living in the wild under the right conditions (feral pigs in woodland ecosystems, feral horses in the open rangeland of the American West, feral dogs and cats where there are opportunities to scavenge at the fringes of human habitation), still, circumstances have forever blocked a true return to their original state. For one thing, the same environmental transformations that pushed out hunting as a way of life have largely pushed out living in the wild as a way of life for the species that gave rise to domesticated animals. Wolves occupy the merest fraction of their original range; wild sheep and goats are restricted to craggy mountain ranges in Siberia, western North America, and slivers of Europe and central Asia; the wild ancestors of cattle and horses long ago vanished from Europe and southwest Asia.

Domesticated animals, like domesticated plants, have also to a great extent lost their adaptations to the wild. Instincts of territoriality and dominance are diminished. Physical defense mechanisms have atrophied; the reduction of flight muscles in domestic fowl are an obvious example.

Generations of captive breeding beyond the initial neotenic changes that established domestic species have only reinforced this helplessness and dependence. A horse that does not know how to compete successfully for a mate or mount a female is not fit for life in the wild. To the extent that such behavior has a genetic component, those traits will be rigorously and swiftly selected against. But in captivity, an animal with otherwise desirable traits will be encouraged to mate with every method at human disposal; the result may be swift racehorses that are utterly incompetent in the art of reproduction. So long as humans are at

hand to assist, such a breed will flourish; on their own, they
would vanish in a flash.

The increasingly widespread use of incubators to hatch
the eggs of domestic fowl has induced a similar epidemic of
reproductive incompetence. Without selective pressure in
favor of hens that go broody and sit patiently on a clutch
of eggs (and in fact with pressures that work considerably
the other way, since a broody hen is not a laying hen, and
laying hens are what humans have selected for), most hens
of modern stock seem not to have the slightest interest in
sitting on nests.

Ewes that benefit from valiant human efforts to save
lambs that, with exasperating frequency, come out wrong
end first or tangled with their twin or triplet brothers and
sisters in the uterus, pass on this propensity to their offspring. Some selection still operates against this among
range flocks that are expected to lamb largely unassisted;
those ewes that have lambing complications simply die. But
in farm flocks the proportion of problem births is growing
steadily, since the "problem" lambs are usually the very
ones we want to save for breeding: The very traits that
make them troublesome to their mothers—the propensity
to conceive twins or triplets, or to grow large lambs—are
both heritable and economically desirable. It can take only
a few generations of such unnatural selection to produce a
flock in which the majority of ewes need assistance at
lambing time. And it takes a rigorous and conscientious
program of culling to prevent it. Feral populations of sheep
in New Zealand that were studied were observed to decline in population, at least initially, mainly because of poor
maternal behavior. The ewes would often have their lambs
out in the open instead of in a sheltered spot, for example.
Good mothering ability is one trait that sheep breeders are

always trying to consciously select for, but the very nature of the farm enterprise—the relative helplessness of the animals and the interest the farmer has in their survival—all but ensures that self-sufficiency on the part of the animals is continually being selected against.

An apt analogy may be in the growing number of people who wear glasses. Presumably, in the bad old days, natural selection tended to disfavor people who were born with a genetic predisposition to nearsightedness; they would have been more likely to walk off cliffs, or stumble into ambushes, or not notice the lion crouching behind the rock than their more able-bodied counterparts. The invention of eyeglasses, despite the obvious benefits to those individuals who are nearsighted, has all but eliminated that selective force; now even those who are likely to bump into walls if they are not wearing glasses have a reasonable shot at surviving to reproductive age and passing on their defective genes to the next generation.

There is an interesting moral analogy, too, in this situation: Everyone who wears glasses agrees that they are a crutch, a nuisance, and a constant reminder of a physical weakness; no one recommends, however, that, to correct this creeping degeneracy in the human species we outlaw glasses. Likewise, one may argue that domesticated animals are degenerates that through dependency and excess kindness from humans have become weak and ever more dependent on the crutch of human care. But calling them "degenerates" does not somehow mean they are less worthy of our consideration. If anything, their degeneracy, which we had a hand in (though were not the original authors of) argues for an even greater responsibility on our part.

Looking at the dependence of domesticated plants and

animals in much broader terms, Coppinger suggests in fact that evolution and time are now on the side of the "degenerates." Dependence not only means there is no turning back; dependence has actually become such a powerful evolutionary force unto itself that it is ushering in a new evolutionary age—what he terms an "Age of Interdependent Forms." In this new age, the dominance of domestic symbioses in the global ecosystem will inevitably lead to a mass extinction of more specialized and independent species as great as that which struck the earth seventy million years ago with the vanishing of the dinosaurs. "Current biological concepts of what is fit and adaptive are at least 15,000 years out of date," Coppinger argues in "The Domestication of Evolution." He continues:

> The fittest strategy of the future may be a system of more cooperative, interdependent relationships among species that are more efficient and more reproductive than the highly specialized self-sufficient competitors which we used to imagine were nature's fittest. The King of Beasts will have been outcompeted by the cooperative strategy of the house cat. And the bald eagle, if it is to survive, may become dependent on human garbage dumps, as a flourishing group is recently reported to have become in the Aleutians. The period in which dinosaurs shook the Earth in search of flesh or flora may seem no more strange to our great-great-grandchildren than the world of self-sufficient, highly specialized "dinosaurs" of our day, stalking prey or marking their territories.

The essential point, as a glance at the population growth of man and his fellow degenerates over the course of recent centuries readily substantiates, is that species outside the dependent alliance simply are incapable of reproducing as swiftly. In 1860, man and domestic species accounted for 5 percent of terrestrial biomass—the net total weight of all plant and animal life on the continents. Today the figure is approximately 20 percent; Coppinger estimates that by 2020 it will be 40 percent, and, if world population reaches twelve billion, as seems possible before it levels off, the figure will be 60 percent sometime in the middle of the next century.

All of which, as Coppinger emphatically points out in "The Domestication of Evolution," is not necessarily the environmental disaster that is so often depicted by those who point with alarm to the extinction of species and the appropriation of ever more of the earth's resources by humans. Change, to an evolutionist, is not good or bad, Coppinger says; it is simply inevitable. The dominance of man and domesticated animals and plants is not the end of nature; it is no more than the beginning of yet another new period in nature's long and varied history. The same evolutionary forces that have brought forth new species, our own included, have eliminated species as well. New ecological balances have repeatedly replaced old as mass extinction has followed mass extinction.

Thus the effort to save endangered species may require much more than just curbing the "excesses of thoughtless polluters," Coppinger writes; it may require outmaneuvering the momentum of evolution itself. He suggests, perhaps slightly facetiously, that one way would be to apply the lesson of domestication and systematically attempt to breed more dependent forms of the species we would save, selecting for neotenic traits. But given the deeply ingrained, romantic image within the environmental movement of nature as something eternal and separate from man, he concedes that that is probably a losing proposition: "Some might rather let the bald eagle go extinct than see it become, let alone aid it in becoming, interdependent enough to share our living space."

The dependence of domestic plants and animals upon humans is part of something much larger than us or them. In a moral sense, it represents a responsibility that has been thrust upon us that it is too late for us to renounce. In an evolutionary sense, it may be something that we could not renounce even if we tried.

VII

HUMAN PIETIES

*He perceived more clearly the cruelty of
Nature, to whom our refinement and
piety are but as bubbles, hurrying
downwards on the turbid waters. They
break, and the stream continues.*
—E. M. FORSTER, The Longest Journey

The image of nature as a peaceable kingdom populated by noble beasts is a recent invention. Only those whose work and day-to-day life took them away from the natural world could cease to see nature as the enemy, the fearful place that threatened life and prosperity at every turn, and see it instead as a sanctuary, a haven of peace from the industrial battlefields of man. The romanticism of today's urban population toward nature and toward those simple savages or farmers who they imagine once lived there in perfect harmony is mystifying to those who continue to make their living from the land—those who see too clearly the cruelties and ambiguities in nature.

The urge to romanticize nature and preindustrial civilization is part of a larger, antitechnological world view that has gained ascendency in reaction to the overblown promises of the Industrial Revolution. Having once exaggerated the promise of technology—electricity too cheap to meter, the elimination of poverty and disease, even an end to war—we are now conditioned to exaggerate its failures, and the environmental movement has given voice and legitimacy to this dissatisfaction. In any battle, there is a temptation to see one's opponent as not just evil but as the source of all evil in the world, and in the fight to curb the excesses of industrial pollution the environmental movement has succumbed to the the temptation. Michael W. Fox, a widely published animal rights advocate (a veterinarian, Fox also writes a nationally syndicated "Animal Doctor" column and the "Pet Life" column for *McCall's* magazine), sees everything from heart disease to suburban tract houses to corrupt government officials as the result of our having violated our covenant with nature. He writes in *Returning to Eden:*

> The promise of a great "marriage" between man and nature, with mankind as the steward of the biosphere, was latent in our hunting stage. It began to blossom when we became farmers and husbanded parts of nature. But now this marriage between man and nature is being destroyed, as are our relationships with each other and with our fellow animals. We are becoming com-

petitors and adversaries. Nations compete and go to war over vital minerals and other natural resources. We no longer lovingly husband the land or our livestock; instead we manage "agribusinesses." ... It was once thought that science and technology would remedy the social ills of humanity by mastering nature. Instead, they have only compounded and increased those ills.

Ecologist and TV personality Paul Ehrlich (who warns in *The Population Explosion* that human population, affluence, and technology are destroying nature) is so intent on blaming our woes on technology that he can find no good even in technologies that will clearly reduce man's "appropriation" of the earth's resources. If biotechnology succeeds in producing natural vanilla-flavoring in the laboratory, he argues, it will still be bad, because it will put out of business seventy thousand vanilla bean farmers in Madagascar. It is a curious argument, since Ehrlichs' proposed remedy for the earth's ills—reducing the world's population, especially in affluent countries—will just as surely reduce demand for vanilla beans.*

Some environmentalists even seem to take genuine satisfaction in technological failure, if for no other reason than for its value as a recruiting tool. One leading animal rights philosopher engaged in a bit of technological

---

* Here as elsewhere, Ehrlich argues that technology will not solve our problems, but even if it does it will be bad because it will strengthen the "control of large corporations." Fox, in his book *Inhumane Society*, says the meat industry is "just one segment of the American agribusiness food production system that violates others' rights in its monopolistic game of control." It is not hard to find the political motive behind these positions.

Schadenfreude over the explosion of the space shuttle *Challenger,* which killed seven crew members: "Animal rights comes at a time," the sage told the *Boston Globe,* "of dramatic, massive failure of the whole rational-mechanistic-industrial structure embodied by the *Challenger* disaster."

It is tempting to point out that only as a result of the whole rational-mechanistic-industrial structure is our society well-off enough to allow people like Fox and Ehrlich to make a living writing books and columns and making TV appearances instead of running around the forest grubbing for tubers. But the more important point is that the premise they are selling—that nature would be fine if only man would leave it alone, and we know that man can leave it alone because that's what he did in the good old days—is hard to square with archaeological, biological, or anthropological evidence. That premise sets a false standard, or at the very least one that is ten thousand years out of date, but it is the basis for much of the success that the animal rights movement has had in advancing its agenda of halting animal research, hunting, and farming.

With Eden established as the reference point for nature and the noble savage the standard for human conduct, the animal rightists have a fairly easy job of horrifying the rest of us—or at least those of us who don't know much about biology, ecology, or animal husbandry—with tales of human actions that appear to depart from these totally artificial norms. The horror story is a staple of the animal rights literature. The animals that star in these tales are almost always furry, and the cruelty they suffer invariably involves an extra bizarre or perverse twist. At times, the animal rightists have to strain a bit to find that extra twist that is supposed to add indignation to the horror, as in the account that appeared in the June 1987 issue of *The Animals' Agenda* of the "rescue" of twenty-five laying hens just six hours before they were to be shipped off to slaughter. The birds, the writer tells us, were to be "sold as low-grade meat for use in soup and pot pies."

Sometimes, the stories aren't even true, but appeal effectively to a vague fear or mistrust of technology. A tract from the Farm Animal Reform Movement urging a ban on the raising of veal claims that "their flesh contains excessive residues of penicillin, tetracyclines, and other antibiotics that immunize the consumer's intestinal bacteria and render these crucial life-saving drugs largely ineffective in fighting infectious diseases," a sentence that in terms of sheer density of wrong statements per word is a model of its kind. Farmers deny that antibiotic residues are even present in veal. But even if they are, they are incapable of "immunizing" the consumer's intestinal bacteria and rendering these drugs "ineffective in fighting infectious diseases" for the simple reason that intestinal bacteria, which play an essential role in the digestion process, do not cause infectious diseases. Although some pathogenic organisms such as the bacterium that causes gonorrhea have acquired resistance to penicillin and other antibiotics (but as a direct consequence of the therapeutic use of these drugs in infected humans, which has the effect of selectively weeding out strains of bacteria that are susceptible to the drugs, leaving only the resistant strains), these drugs overall remain astonishingly effective a half century after their introduction. A huge survey of forty-three million tests for resistance made in hospitals from 1971 to 1982 found no overall increase in resistance to fifteen different antibiotics over this period, according to researchers Barbara Atkinson and Victor Lorian.*

Or consider the episode that took place in August 1987, when members of the Animal Liberation Front broke into a laboratory at the Department of Agriculture's Animal Parasitology Institute in Beltsville, Maryland, and stole twenty-five cats. Ingrid Newkirk of the People for the Ethical Treatment of Animals, in announcing the action, said the cats, which were infected with the parasite that causes toxoplasmosis, were going to be left untreated to suffer or die in the lab, and had been infected by being force-fed infected mouse brains through a stomach tube. It made a good story. It was also a total fabrication. The cats were infected, but toxoplasmosis, which can cause blindness and brain damage in human babies born to mothers who are first exposed to the disease during pregnancy, causes no symptoms at all in most cats. Cats are the primary carriers of the disease, and in fact nearly all cats that are allowed to roam outdoors become infected sooner or later. Even though they rarely get sick, though (and almost never die), infected cats do transmit the disease to humans when they pass the parasite in their feces into litter boxes, flower beds, or other places that humans come in contact with. Toxoplasmosis also causes abortions in sheep, goats, and pigs. In humans whose immune systems are compro-

* Some scientists have raised concerns that the routine, low-level administration of antibiotics to farm animals increases the populations of resistant strains of bacteria in those animals' guts, and that this resistance might be transferred by bits of DNA called plasmids to other bacteria, including those responsible for human diseases. Such a transfer to human pathogens has never been observed, however.

mised, such as AIDS patients, the parasite can cause brain damage and, not infrequently, death.

Many of the horror stories, though, especially the ones about farming practices, rely for their effect upon the general public on the growing squeamishness of people who are far removed not only from animals, but from death and serious illness in their own lives. Tail docking of lambs can certainly be made to sound rather grisly. Even the least traumatic method, which involves putting a small rubber ring around the tail when the lamb is a few days old, clearly causes the animal some discomfort for fifteen or twenty minutes. (After that, the tail becomes numb; it drops off after a couple of weeks.) But leaving the tails on often results in a much grislier alternative, as anyone who has seen a dung-covered tail infested with maggots could attest.

When animal rightists decry the stress created when chickens are confined to cages, it is hard not to sympathize. When we see the pictures of birds with bare spots on their skin, their feathers pecked out by other "highly stressed" birds, we are easily persuaded that this is unnatural, surely not the way things would be in the peaceable kingdom. I have seen one of my well-fed, low-stress, "free-range" hens (to use the trendy terminology for a barnyard chicken) run the length of the barn to peck another hen. Most people have not.

The burden of all of these stories is that man is committing not only an atrocity against animals but an atrocity against nature. He is seeking to control what he should not, to arrogantly expand his dominion, to enslave what would otherwise be free. (Some animal rightists even object to shearing and ear-tagging sheep, since these limit the "freedom" of the animal.) As Coppinger puts it, this view sees domestication of animals as "but another artificial human invention that is somehow ruining nature."

In truth, it is nature that has "ruined" nature, if by "nature" we mean the idealized place of the urbanite's dream.

Biological studies tell of a world of continual struggle, disease, and stress. One of the best examples comes from an examination of hormonal responses and behavior in wild olive baboons in the Masai Mara National Reserve in Kenya. "Food is plentiful ... predators are few, and infant mortality is low," explains researcher Robert Sapolsky. A Garden of Eden? Well, not exactly: "With the luxury of plentiful resources and free time, the animals can devote themselves to distressing one another. . . . Violence itself is actually rare, but the hint of violence is ever present." The males form a dominance hierarchy that bears a striking resemblance to the social structure of a typical junior high school, and life for a subordinate male is one of continual bullying. "Subordinate males may laboriously dig tubers from the ground only to have the food nonchalantly seized by dominant males," writes Sapolsky. "Dominant males who lose a fight often seek a subordinate on whom to vent their frustration, and they are likely to displace aggression on the innocent bystander without warning."

As Sapolsky points out, the stress response in all animals is a survival mechanism, preparing the animal for fight or flight. Hormones released in response to stress cause glucose to be mobilized from storage and blood diverted to those parts of the body needed for a quick physical response—the heart, the brain, and the muscles—and away from such "nonessential" functions as growth, digestion, disease resistance, and reproduction. But continual activation of these hormones can lead to chronic physical disabilities, including hypertension, ulcers, impaired growth, and decreased fertility. By actually measuring the hormone levels of wild baboons, Sapolsky found that subordinate males have chronically high levels of stress hormones in their blood even when at rest. And there is some evidence that, just as in humans, this chronic stress is taking a physical toll. Compared to the dominant males, the subordinate animals had lower levels of disease-fighting white blood cells and lower levels of the "good" form of cholesterol that helps prevent heart disease.

The biological reality of parasitism is another uncomfortable fact of nature that few of us ever are forced to encounter. Virtually every animal in the wild, especially those we see as paradigms of nature at its most majestic, such as the lion and the wolf, is full of parasites. A little study of the subject goes a long way toward changing one's ideas of the benignity of the natural world. To be more objective about it, parasitism is a prime illustration of the competition between species that is inherent in evolution. Sheep, for instance, are host to at least eight different species of roundworms, six tapeworms, one flatworm, and one especially gruesome parasite known as the lungworm. Sheep ingest the larvae of the lungworm while grazing. When the larvae reach the small intestine, they burrow through the bowel wall into the lymphatic system, follow that to the heart, where they enter the pulmonary blood vessels, emerging finally into the lungs. There they develop into a mass of threadlike adult worms, which begin to slowly suffocate the animal. Affected sheep become weak; they pant in a struggle to get enough air; eventually they either cough out the worms or die.

The award for the most gruesome parasite could go to a marine organism known as the hagfish, which bores its way into a fish's intestine and literally eats its host alive from the inside out.

Some animal rightists, pressing their case that farming is inherently stressful and unnatural, claim that parasitism is only a problem when animals are artificially crowded together. It is a nice attempt to shift the blame to man for parts of nature that are decidedly ugly, but it just does not stand up to scrutiny. Parasites owe their existence not to man but to evolution. A thirty-three-thousand-year-old specimen of fossilized horse dung found in Siberia—predating the domestication of the horse by about thirty millennia—contained ova of the parasitic worm *Strongylus edentatus,* still a serious problem for horsekeepers today.

It is interesting that many of the most popular wild animals these days are predators—wolves, coyotes, grizzly bears, eagles. One wonders how popular they would be if everyone was well acquainted with a gory publication produced by the Texas Agricultural Extension Service, *Procedures for Evaluating Predation on Livestock and Wildlife,* which comes complete with full-color pictures to help farmers identify the source of predator losses. Here is how the coyote leaves its mark:

> Some coyotes kill by attacking the flanks or hindquarters, causing shock and loss of blood. This is quite common on calves, but is less common with sheep and goats. It does seem to occur more often in sheep during the winter months, possibly because of heavy fleece during this period. Death of the calf and severe injuries to the genital organs and hindquarters of cows are characteristic when coyotes attack cows giving birth. . . . Multiple coyote kills are frequent and many of these kills are not fed upon.

The publication also informs us that it is quite common for black bears to feed on the udders of lactating ewes, sometimes leaving the animals to die what is surely an extremely painful and horrible death.

The determination on the part of environmentalists and animal rightists to romanticize the perfection of nature and to ignore its obvious cruelties echoes efforts made a century ago by Christian theologians to reconcile the difficult facts that (1) God created the earth and all life thereon; (2) God is perfect in his judgment and his mercy; and (3) diseases, as biologists had begun to demonstrate, are caused by other living beings, presumably just as much a part of God's creation—the only difference being that their sole purpose in life is to inflict suffering in exceptionally unpleasant ways. Mark Twain, for one, clearly despised the hypocrisy of the preachers who regularly claimed as God's work science's eradication of a disease; an objective assessment of nature, Twain said, could only lead one to the conclusion that flies, parasitic worms, and microbes were far and away the Creator's favorites. Twain, in his *Letters from the Earth,* an extended satire never published in his lifetime, described the scene of Noah's ark:

> Noah and his family ... were saved, yes, but they were not comfortable, for they were full of microbes. Full to the eyebrows; fat with them, obese with them; distended like balloons. It was a disagreeable condition, but it could not be helped, because enough microbes had to be saved to supply the future races of men with desolating diseases, and there were but eight people on board to serve as hotels for them.... There were typhoid germs, and cholera germs, and hydrophobia germ, and lockjaw germs, and consumption germs, and

black-plague germs, and some hundreds of other aris-
tocrats, specially precious creations, golden bearers of
God's love to man, blessed gifts of the infatuated Father
to his children—all of which had to be sumptuously
housed and richly entertained. . . . The great intestine
was the favorite resort. There they gathered, by countless billions, and worked, and fed, and squirmed, and
sang hymns of praise and thanksgiving; and at night
when it was quiet you could hear the soft murmur of it.
The large intestine was in effect their heaven. They
stuffed it solid; they made it as rigid as a coil of gaspipe.
They took pride in this. Their principal hymn made
gratified reference to it: "Constipation, O constipation/
The joyful sound proclaim/Till man's remotest entrail /
Shall praise its Maker's name."

There is a difference, of course: The nineteenth-century
theologians were trying to find in nature a reflection of
God's perfection, while today's nature-worshipers are seeking to show nature as perfect in its own right. But the
striking parallel is the determination of both to steer clear
of troubling complications.

The proposition that primitive societies lived in worshipful respect of nature is an essential corollary to the
proposition of nature's perfection, for it implies that we can
regain this lost Eden merely by abandoning the technological excesses of recent days. The notion that hunter-
gatherers and primitive agriculturists had attained in their
relationship with the earth a spiritual high ground that
modern man has lost is widely accepted, and not just by
animal rightists. The well-known religious beliefs of the
American Indians lend much credence to the idea. "The
bison was sacred and gave us both food and shelter," explained the medicine man Black Elk of the Oglala Sioux.

Archaeologists have long gone along with this view as well. "There has been a real undercurrent in archaeology that prehistoric man was an ecological hero who could walk through the forest without snapping a twig," says Julio Betancourt, a researcher at the U.S. Geological Survey. But recent research by Betancourt and others is forcing many archaeologists to question this assumption. The archaeological record, as it now turns out, is a record replete with entries telling of extinction and destruction at the hands of man. The extinction of the mammoth in America eleven thousand years ago, for example, now appears to have been the direct consequence of hunting by Paleo-Indians, the same people whom animal rights activist Michael W. Fox is forever holding up as examples of how we can live with worshipful respect for Mother Earth.

Other primitive societies managed to totally transform local ecosystems. Betancourt has put together a convincing case of a man-made environmental disaster engineered by the pueblo-dwelling Anasazi Indians eight hundred years ago. The Anasazi, who lived in what is now New Mexico and Arizona, built an elaborate complex of roads, irrigation channels, and five-story stone and wood-beam pueblos, some containing as many as eight hundred rooms. All were abruptly abandoned around A.D. 1200.

At Chaco Canyon, a major Anasazi site in New Mexico that is now a barren desert, Betancourt found that the Anasazi had systematically stripped bare the pinyon pine and juniper forest that had provided firewood and construction beams. The ancient refuse heaps of pack rats have preserved an accurate historical record of this environmental disaster. The heaps contain leaves, twigs, and other odds and ends collected within a short distance of the rats' home burrows; glued together with the rats' urine and sheltered

below ground from the weather, they provide a time cap-
sule of local vegetation. Radiocarbon dating was used to
determine their age. The pack rat heaps contain an abundance of pinyon needles and juniper twigs—until A.D. 1200,
that is. At that point all traces of juniper and pinyon
promptly disappear. Electron micrographs and tree-ring
dating of the logs used in the Anasazi pueblos show that by
around A.D. 1030 the Indians had been forced to shift to a
new species of wood, Douglas fir, which grows only on
mountaintops at least fifty miles distant from the site. They
kept up a brave effort for several decades, dragging logs
these huge distances by hand, but finally abandoned the
site.

Archaeologists had long assumed that the abandonment
of Chaco Canyon was the result of a climatic change, which
presumably also explained the apparent mystery of why the
Indians would have built a city in the middle of what is
now a desert. Betancourt's work, however, makes it clear
that the environmental disaster that befell the Anasazi was
largely self-inflicted. Very simply, a primitive society
turned a rich forest into a desert through overharvest-
ing—an ecological catastrophe that remains a millennium
later.

Then there is the case of the Polynesians. Studies of
fossil birds of Hawaii and the South Pacific by Smithsonian
biologist Storrs Olson has uncovered "one of the swiftest
and most profound biological catastrophes in the history of
the world." It was not, as one might think, caused by the
arrival of Europeans with guns. By the time Captain Cook
passed through in the eighteenth century, 80 to 90 percent
of all of the species of birds of the region had already been
wiped out. On Hawaii, fossils of more than fifty species of
birds that are today extinct have been unearthed. As the

Polynesians spread throughout the region centuries before, they brought with them dogs, pigs, and rats (a particularly large ecological jolt for islands such as Hawaii that had no native mammals other than bats) that raided the nests of ground-dwelling birds. The Polynesians cleared land for farming. They hunted. They brought domestic chickens that may have spread diseases such as avian malaria. Wherever Polynesian artifacts appear in the archaeological record, a whole range of birds, including parrots, pigeons, and flightless geese, simultaneously vanish. Some of these species survive on a few remote islands, but others live on only in the legends of the islanders, who have names for birds they have never seen; their descriptions of these legendary birds precisely match the fossil birds that archaeologists have unearthed.

Finally, to dispose of the myth that ancient peoples were all noble savages who could teach us a thing or two, it's worth considering the case of that noble people the Yanomamo, who live in the rain forests of the Amazon. Napoleon Chagnon, an anthropologist at the University of California-Santa Barbara who studied this hunter-gatherer culture for twenty-three years, concluded that they were among the most violent societies on the planet:

> Although there are customs and general rules about proper behavior, individuals violate them regularly when it seems in their interest to do so. When conflicts emerge each individual must rely on his own skills and coercive abilities and the support of close kin. Most fights begin over sexual issues: infidelity and suspicion of infidelity, attempts to seduce another man's wife, sexual jealousy, forcible appropriation of women from visiting groups, failure to give a promised girl in marriage, and (rarely) rape. . . . Approximately 30 percent of

adult male deaths are due to violence ... nearly 70 percent of adults over an estimated 40 years of age have lost a close genetic relative due to violence....

The Yanomamo are frank about vengeance as a legitimate motive for killing.... It is dangerous to provoke a grieving person no matter what the cause of death of the lost kin. It is common to hear statements such as "If my sick mother dies, I will kill some people."

Nearly half the males over age twenty-five have participated in a killing. The weapons of choice are bows and arrows and clubs.

Lest anyone think this is an aberration, pathological studies of archaeological specimens show not only the obvious fact that primitive man suffered from a variety of debilitating ailments, including osteoarthritis, tuberculosis of the spine, scurvy, rickets, and osteomyelitis (a particularly painful infection of the bone), and had a much shorter life span than modern humans (the average life expectancy for someone born in an early agriculture culture in Cyprus, for example, was twenty-two years), but that many met their ends at the hands of their fellow man. Jessie Dobson writes in *Diseases in Ancient Man:*

What is more depressing, however, is the continuing evidence of injury through violence persisting through the centuries. Neolithic peoples are usually considered to have been peaceful peasants, but a study of their remains shows plainly that this was not so. A Neanderthal skull shows a scar above the right eye socket and there was also a healed fracture of the cranium. Cromagnon man shows a perforating lesion in the pelvis, possibly a stab wound. In Iron Age skulls there occur oblong apertures in the frontal region, the result of axe or sword blows.

As for the loving stewards of the land, their form of agriculture was sustainable not because of any innate wisdom but only because it was practiced on such a small scale. Slash and burn was not invented by "agribusiness"; it was being practiced in Europe by 7000 B.P. to clear forests for growing grain, and was also the method of cultivation preferred by American Indians. The traditional farming practices in other places and at other times have been little better. In *Night Comes to the Cumberlands,* Harry Caudill tells the story of the agricultural extension agent in the 1920s who tried in vain to convince local farmers to abandon their totally traditional and totally exploitive brand of farming that had led to massive erosion of the soil. One Kentucky farmer listened with mounting skepticism to the agent's recommendations: Plant soil-conserving cover crops, sow legumes that add nitrogen to the soil, stop plowing steep hillsides. Finally, the farmer had had enough and retorted: "I don't see how you know so much about farming. After all, I've wore out two good farms on this creek since you was borned and I bet you ain't even wore out one yit!"

Raising all of this evidence about the cruelty of nature and the environmental depredations committed by traditional societies is not to argue that because nature is cruel we can be cruel, or that because our ancestors were environmental boors we can be; far from it. But neither can we create an artificial standard, based on a romantic image of nature as perfect and peaceful and of primitive man as an ecological saint who could live his life without harming another living thing. The struggle between species is a grim reality of the world, and the evolutionary advantages that led to the "domestic alliance," Coppinger's term for the ecological relationship between man and domesticated animals, underscore some genuine improvements in the lives of species that cast their lot with man's. Freedom from predators, from starvation, and from parasites are not advantages to be dismissed casually.

In *Animal Liberation,* the animal rights philosopher Peter Singer sidesteps these complications by sticking to a purely moral argument against farming. Just as defenders of slavery tried to justify their immoral conduct by citing the improvement in the lot of "inferior"Africans brought to America, writes Singer, so defenders of animal agriculture try to justify their exploitation of animals with similar reasoning. Singer's contention is thus that it does not matter how good conditions are on the farm; by keeping animals at all we are depriving them of a basic right to freedom. It is an argument that even those who do not buy into the animal rightists' agenda often find troubling and difficult to answer. Even "the steady supply of food on a farm is not an unmitigated blessing," says Singer, "since it deprives the animal of its most basic natural activity, the search for food. The result is a life of utter boredom.... Surely the life of freedom is to be preferred."

Again, this is an argument that is best answered by nature herself: The evolutionary changes that led to domestication occurred precisely because "freedom" was *not* to be "preferred." (Singer's analogy also fails in that the racist arguments used to justify the subjugation of supposedly inferior races were pseudoscience; the line between species and the inevitable competition between them is a biological reality.) Whether a farm animal is "bored" by the prospect of finding a rack of hay in the same spot each evening is probably impossible to know for certain. But even a moderately observant pet owner ought to be suspicious of that argument. Although some feral dogs and cats do manage to survive at least awhile (and in favorable climates) through scavenging and hunting on their own, the juvenile trait of begging for food is a basic part of the behavior of all domesticated animals. On biological grounds, one cannot argue that an animal is deprived by not being allowed to search for food on its own if it no longer has a biological urge or ability *to* search for food.

That this is something fundamental in the evolution of domesticated animals becomes clearer if one contrasts the psychological condition of a pet or a farm animal with that of wild animals kept in similar conditions. Consider two tales of two animals. The first is of "Smokey," a bear at the National Zoo in Washington. Smokey was brought to the zoo as a young bear orphaned in a forest fire; he had essentially never known freedom, and certainly not independence. But even though he had little or no experience searching for food on his own, the instinct was undeniably there. The black bear *(Ursus americanus)* could in fact be the original source for the image of animals as wild and free, the sort of animal whose dignity and even sanity is destroyed in confinement. Black bears spend a considerable amount of time searching for all kinds of food; they have to eat enormous amounts, especially in the fall to build up fat reserves. Although Smokey's pen was far from a barren cage—it was outdoors, roomy, and to some extent a re-creation of natural habitat—the base was concrete (necessary to keep down intestinal parasites, which bears are especially prone to), and he was fed at regular times in a regular place. Under these conditions, Smokey developed a series of what biologists term "stereotypy" behaviors, the most annoying of which was pacing from one end of the pen to the other along precisely the same route in front of his swimming tank, taking precisely seventeen paces, then turning his head to the side, turning around, and pacing back. According to his keepers, he would do this all day long—as much as eight hours per day.

What finally eliminated the stereotypy behavior was a change in feeding practice made possible by the replacement of the concrete base with a more natural cover, following the development of a new anthelmintic drug, Ivermectin, which was able to control the parasite problem.* Keepers then began hiding small amounts of food under leaves and brush and in logs throughout Smokey's pen. The new feeding strategy, which gave Smokey a more natural way to fill his time, reduced his pacing to twenty minutes per day. It is not necessary to indulge in anthropomorphism to conclude that the bear was indeed bored having his food placed in front of him, bored to the point that he developed a psychological abnormality. And, given the bear's limited experience in the wild, one can reject the idea that any animal at all can adapt to a life of confinement and hand-feeding merely by training. Clearly, there is something fundamental about the behavioral makeup of the black bear that renders it ill suited to a life of being cared for.

* Animal rightists frequently argue that animal research is not only cruel but unnecessary and even scientifically unsound since animals respond so differently from humans; the development of Ivermectin is a persuasive counterexample. The drug was developed and tested extensively in animals, and is now among the most effective horse, pig, cattle, and sheep wormers; it is used as a one-a-month pill to prevent heartworm in dogs, replacing the previous drug, which had to be given daily; and it is the only effective treatment for one of the most hideous of human parasitic diseases, river blindness, or onchocerciasis. An estimated ten million people in Africa and Central America have suffered serious eye damage or total blindness from the disease; an estimated forty million are infected.

The second story that illustrates the danger of applying moral concepts such as freedom to the circumstances of animals concerns a sheep in my barn, who developed an awful condition known as vaginal prolapse. This is a problem that can afflict ewes carrying twins or triplets who are eating a lot of hay to meet the increased nutritional requirements of pregnancy. The increased pressure on the cervix from the lambs and the digestive organs forces the vagina to evert out of the vulva. If left unattended, the exposed vagina can dry out and crack, or even rupture, causing death of the ewe and lambs. But more immediately, the prolapse blocks the urinary tract and can cause the bladder to burst. As with so much of sheep obstetrics, it constitutes a genuine veterinary emergency.

The ewe was a yearling, carrying twins. The usual treatment is to turn the ewe upside down with its hind end propped against a straw bale for support, wash the prolapse off thoroughly with soap and water, and gently reinsert the whole mess, and then, since a prolapse will inevitably recur, insert into the vagina a plastic paddle known as a bearing retainer, which is tied in place to the wool on the sheep's hips until lambing time.

In this case the usual treatment didn't work. The next attempt involved suturing the vulva closed, an operation carried out at nine o'clock one evening with dental floss and a carpet needle sterilized in rubbing alcohol. This held for about a day, at which point we called in the vet, who applied much fancier stitching. That didn't work for long, either.

Finally, it was clear that the ewe would die unless drastic measures were taken. The only option was to place the ewe in a stanchion crate elevated at the rear, so that she would always be tilted forward at about a five-degree slope, and gravity would keep everything in. For about a month, until she gave birth, the ewe was confined for hours on end in a box no bigger than herself, her head held in a wooden slot, her feed and water placed in front of her. She could stand up and lie down, but not stretch, turn around, or graze. Farmers can be as anthropomorphic as anyone else, and I was sure that this would prove intolerable to an animal used to freedom. The first two times we put her into the crate it was indeed a struggle. But to my astonishment, the ewe adapted to the new life almost instantly. We would let her out into a paddock for a few minutes each day to walk around, but once she was back in the barn she would walk, on her own, right into the crate, stick her head into the stanchion, and wait to be fastened in. That was the place she was fed, and clearly the attraction of food outweighed the loss of "freedom." This was an animal that was the exact opposite of Smokey. Despite a life of freedom in which she was left to find her own food—for two thirds of the year our sheep get almost all their food from pasture— she was by her very nature willing to accept confinement and "boredom," so long as it was accompanied by regular feeding.

There is no way to ask an animal if it is bored, but what we can know with some certainty is that the systematic neoteny that has occurred in all domesticated animals has rendered them *genetically* much more dependent than the wild animals that Singer has as his model when he speaks of animals' "natural" activities. They are not merely dependent because we choose to confine them to barns and fields and feed them; they are dependent by their evolutionary history.

And despite our anthropomorphic preconceptions of what constitutes boredom or freedom to an animal, the animals often have their own ideas. In an experiment designed to see whether laying hens preferred life in a cage or in a garden run, hens were very simply offered a choice. The result, which would come as absolutely no surprise to anyone who has worked with animals, is that the hens showed a strong preference for the setting they were used to. Hens that had been confined to cages chose to go into the cages; hens that had been brought up wandering about preferred the garden run.

It may make for a smart philosophical debating point to draw analogies to human slavery, and to assert that because freedom is a fundamental human yearning, and because humans prefer freedom to slavery even when freedom comes at a price, that the same must hold true for domesticated animals. But there is a maddening intellectual arrogance inherent in the determination of Singer and other animal rights philosophers to ignore a century of animal behavioral studies that could actually answer some of these questions. Experiments can offer considerable insight into the innate behavior of domesticated animals and allow us to draw some distinctions between what animals actually require for their well-being as opposed to what we merely imagine they require.

Experiments that offer animals a choice do have their pitfalls, to be sure. The mere fact that a dog will prefer a steak to a bowl of dried dog food does not necessarily mean we are being cruel or that the animal is in any way suffering if we feed him nothing but dry dog food. But such experiments can disabuse us of some of our more egregious anthropomorphic tendencies, as happened several years ago in Great Britain when an official government committee was charged with examining the welfare of animals raised under intensive farming conditions. The committee, with due deliberation, recommended that cages for laying hens be constructed with floors made not of lightweight hexagonal chicken wire, which was the prevailing practice, but of a stronger rectangular mesh, which they believed would be significantly more comfortable for the birds. Producers were opposed, since the heavier wire cost considerably more and because in any case having to replace all of their existing cages would be an expensive proposition. The rectangular mesh also tended to result in more cracked and broken eggs. Finally, somebody had the wit to ask the chickens what they preferred. A simple study was carried out; hens were allowed to move freely between the two types of floors. They voted overwhelmingly for the hexagonal wire.

If animals so often defy human preconceptions, nature defies human moralizing. Given man's place in the ecosystem, one can scrupulously adhere to the animal rightists' moral creed, and reject meat, fur, leather shoes, pharmaceuticals developed using laboratory animals, and springloaded mousetraps, and still be responsible for a pile of animal corpses. And that is the fundamental problem with the moral argument for animal rights: It rather overlooks the way nature operates. The human ethic that says simply "do not harm" is inadequate when it comes to the interconnections and conflicts of nature—or rather, it misses the point.

Consider the obvious good deed of adopting a cat from the local animal shelter. Here is a clear-cut choice between life and death, between kindness and neglect. No animal rights activist would fail to applaud. But a cat introduced into a neighborhood is not being introduced into a vacuum. It is becoming a part of a complex ecosystem, and affects it, not necessarily for the better. Two British scientists, Peter

Churcher and John Lawton, their curiosity aroused by the astonishing number of bird, mouse, and rabbit corpses brought home by their otherwise well-fed cats, decided to undertake an intensive study of the ecological impact of cats in one village in Bedfordshire. Their method was straightforward: They simply asked the villagers to collect all the prey brought home by their pets during a twelve-month period. By year's end, the total take of the village's 78 cats reached 1,100 animals. The 130 house sparrows bagged by the cats amounted to one third of all the sparrows tallied in the village at the start of the breeding season. The actual total may have been much higher: A study of farm cats in Illinois found that cats bring home only about half their catch (they eat the rest on the spot or abandon it to scavengers). Churcher and Lawton calculate that the 5 million cats in Great Britain are responsible for the deaths of about 70 million sentient creatures a year. In the United States, the figure would be closer to 1 *billion*.

Animal rightists often assert that while it is true that nonhuman carnivores kill, they do so only out of necessity; only man kills for pleasure. Interestingly, Churcher and Lawton found that the amount a cat was fed at home seemed to make no difference in its pursuit of prey. "Well-fed and apparently contented cats are often ruthless killers," they noted. Watching my barn cat play with a clawed and bitten but not quite dead mouse also makes me question the idea that nonhuman carnivores are always nobly purposeful and sparing in their work. My cat will typically grab the mouse in its mouth, then immediately let it go. It will repeat this until the mouse is dazed and weakened; then it starts batting the mouse with its paws to try to get it to move so that it can grab it again. The entire process lasts ten minutes or more; I once came upon my cat literally hurling a barely living mouse several feet across the driveway and rushing forward to pounce on it before it had skidded to a stop. By any objective standard, a mouse faces a far more humane death in a spring-loaded trap, and a rabbit at the receiving end of a hunter's bullet.

It is a nasty trick that nature plays, turning an act of seeming compassion into an act of destruction—and vice versa. But it is an inescapable fact of life. "The laws of nature have evolved so that the continuation of each species is ensured, in spite of the destructive forces that appear to be at work when, for example, a wolf pack tears a deer apart ... the deer is adapted from the point of view of population to such predation. It 'expects' to be killed, or it lives to be eaten." That is Michael Fox's argument in *Returning to Eden* for allowing the natural forces of nature to be left to work. If man and domesticated animals are a product of nature and a part of it, that same law surely applies.

VIII

A NATURAL
ETHIC

*George's son had done his work so
thoroughly that he was considered too
good a workman to live, and was, in fact,
taken and tragically shot at twelve o'clock
that same day—another instance of the
untoward fate which so often attends dogs
and other philosophers who follow a train
of reasoning to its logical conclusion, and
attempt perfectly consistent conduct in a
world made up so largely of compromise.*
—THOMAS HARDY,
Far from the Madding Crowd

It is an old and bad joke that hunters love animals,
which is why they go out and kill them. More than a
century ago, Henry David Thoreau was already taking a
jab at the argument commonly heard today that hunters
are the most ardent protectors of wildlife. "Already a
change is taking place" in the popularity of hunting among
boys in New England, he wrote, "owing, not to an increased humanity, but to an increased scarcity of game, for

perhaps the hunter is the greatest friend of the animals hunted, not excepting the Humane Society."

The intent was ironic, yet Thoreau himself understood something about the deeper meaning of hunting and fishing that escapes the modern-day, humorless critics of these activities, to whom they represent nothing more than killing. Despite an uneasiness that grew in him over time with the idea of taking the life of a fellow creature, Thoreau saw hunting as an experience that awakened a self-awareness of man's place in the natural world:

> I found in myself, and still find, an instinct toward a higher, or, as it is named, spiritual life, as do most men, and another toward a primitive rank and savage one, and I reverence them both. I love the wild not less than the good. The wildness and adventure that are in fishing still recommended it to me. I like sometimes to take rank hold on life and spend my day more as the animals do. Perhaps I have owed to this employment and to hunting, when quite young, my closest acquaintance to Nature.... Fisherman, hunters, wood-choppers, and others, spending their lives in the fields and woods, in a peculiar sense a part of Nature themselves, are often in a more favorable mood for observing her, in the intervals of their pursuits, than philosophers or poets even, who approach her with expectation. We are most interested when science reports what those men already know practically or instinctively, for that alone is a true *humanity,* or account of human experience.

It is not just the chance to take a little nature walk and admire the woodchucks and deer that Thoreau is describing. It is the chance to come face-to-face with our part in

nature itself, to live one of those experiences that Aldo
Leopold (himself a hunter with profoundly mixed feelings
about hunting) insisted had value because they remind us
"of our dependency on the soil-plant-animal-man food
chain, and of the fundamental organization of the biota."

The point is not an idle one. It is not simply that an
individual gets to act out a little drama from his primitive
past when he goes hunting in the woods. Leopold has
struck the heart of it when he says how cluttered our
relation to nature has become. A person who buys his
chicken in cut-up pieces wrapped in plastic in a supermar-
ket refrigerator case is the same person who can be con-
vinced that the way to save the elk at Yellowstone is to feed
them hay in the winter, or that docking lambs' tails is cruel
and that there probably ought to be a law to stop people
from committing such abuse.

To people who work closely with animals, there has
never been any contradiction between respecting them,
grasping the essential truth that they are living beings with
a will of their own, and at the same time using them
judiciously for human purposes. "You find among hunting
people all over the world a very intimate, appreciative
relationship to the principal food animal," explained my-
thologist Joseph Campbell. "It is thanking a friend for
cooperating in a mutual relationship." The Kung people of
Africa, basically a late Stone Age hunter-gatherer culture,
for example, ask forgiveness of an animal's spirit after the
kill. They don't try to pretend that there is no ethical cost,
or guilt even, inherent in the act of killing the animal; they
confront it head on.

In Laurens van der Post's African novel *A Story like the
Wind,* there is a scene in which a marauding elephant is

killed. 'Bamuthi, a Matabele chief, solemnly surveys the scene in silence, then turns back to his people and declares: "See that you thank our lord the elephant for allowing himself to be killed so that we can live. See, there in his body we have food for many more days than the maize and millet which he trampled under his great feet could have given us." Van der Post paints a world in which hunting is an inseparable part of an ethos that includes an intimate knowledge of the animal life of the bush and a deep reverence for it. The act of shooting itself is described by Mopani Théron—a hunter by profession who, in his disgust at the ravages of unskilled and ignorant hunters and the disappearance of African game, became a ranger and took up the work of preserving the continent's wildlife—as an almost mystical communion:

> He remembered Mopani teaching him to shoot [narrates the novel's central character, the boy François] and telling him that shooting was not a matter of the will but a kind of two-way traffic between target and rifleman and that if one wanted to shoot accurately without hurt or unnecessary pain to animals, one must never force one's shot by pulling at the trigger with one's finger. Instead one had to keep the gun aimed truly at the target, until the target filled not only all one's eyes but activated one's imagination, until one's finger gradually tightened on the trigger, releasing the shot only when target and rifleman were one. Mopani would talk almost as if shooting for him were allegorical, and the rifle the contemporary version of the symbol the sword once was in mind of medieval man; the image of the spirit in its everlasting battle against falseness and unreality in life.

The battle against falseness and unreality is one that we have lost in the twentieth century. We do not confront nature in its reality, and we are no more civilized for it—only more naive. "We cannot but pity the boy who has never fired a gun," wrote Thoreau; "he is no more humane, while his education has been sadly neglected."

Some scholars who have studied the development of human attitudes toward animals make much of eighteenth- and nineteenth-century accounts—written by founders of the humane movement—describing the depredations committed by farmers who beat their horses to death or whose idea of a good time was baiting a bull with dogs. The implication is that the agriculture culture is by its very nature cruel and exploitive. Those of us brought up on *Black Beauty* tend, at least subconsciously, to buy into this argument. We may not think farmers are actually sadists, but we suspect them of callousness, at least. We wonder how they can raise animals and send them off to slaughter without becoming dulled to their needs and pains while alive. In an essay entitled "The Emergence of Modern Pet-Keeping," Harriet Ritvo, who has written a well-regarded history of Victorian attitudes toward animals, argues that the famously dotty British affection for animals is a very recent phenomenon, the product of a general social reaction against the cruelty that had been the norm. The rise of pet-keeping in the nineteenth century only became possible, she says, once animals had become representatives of "a nature that was no longer threatening" thanks to man's progress in controlling the natural world through science and engineering. "Only at that point could ordinary people interpret the adoption of a representative of the elements (however tame and accommodating) as reassuring evidence of human power, rather than as a troublesome reminder of human vulnerability." Before the mid-nineteenth century, she says, those who kept pets were largely the elite.

The mistake Ritvo and others make, I believe, is to focus too closely on the words of the literate. There is no doubt that cruel practices occurred. There were farmers who beat their horses to death; there were cockfights and dogfights and bull baitings. (In Great Britain, laws were passed prohibiting wanton cruelty to livestock and draft animals in 1822 and outlawing cruel animal sports in 1835.) But to look to the testimony of reformers for an accurate historical measure of how widespread cruelty or cruel attitudes were is asking too much of tainted evidence. A future historian might equally well cite newspaper accounts (or, to make the analogy more accurate, literature put out by handgun-control advocates) of murders in American cities in the 1980s to demonstrate that our culture had no respect for human life. And the mere fact that the widespread keeping of pets as pets is a fairly recent phenomenon in Western society (or that those certain members of the elite who did keep pets in earlier centuries were the object of ridicule by certain others) ignores a world—literally—of evidence about the dominant values of agricultural societies toward the animals they depended upon.

Although the "ordinary" people who have kept dogs for thousands of years may have had utilitarian justification (hunters, guards, herders), the idea that they treated their animals as mere objects (as the animal rights literature so often claims) is very difficult to square with the relationship that has to exist between trainer and dog. Crude cruelty may suffice to produce an effective watchdog whose job is limited to biting anyone within range, but it simply does not work when the goal is anything more refined. Teaching a Border collie to herd sheep, for example, requires patience and empathy as much as it requires firmness. A young pup has to be encouraged to develop his natural instinct to run out and cut off the path of a retreating animal, but at the same time has to be discouraged from carrying this element of hunting behavior any further. Sheep, despite their reputation, can be very intimidating to an inexperienced dog; and too harsh a correction when the dog decides to seize hold of an animal can forever discourage him from facing down an old ewe who turns on the dog, lowers her head, and stamps her foot.

The respect that Joseph Campbell finds in primitive hunter-gatherer societies toward the principal food-animal carried over into many agricultural societies, too. Cattle, which played a central role in the economy of ancient Egypt, are depicted in many friezes and drawings being carefully tended and cared for. They were also the object of religious reverence, as anthropologist Richard Lobban of Rhode Island College notes, notably in the form of one of the most enduring Egyptian goddesses, Hathor, who took the form of a woman with the horns of a cow. Cattle and humans were found buried together in predynastic times, as early as 6000 B.P.; by the time of the Second Dynasty in the Old Kingdom (4000 B.P.), they were being embalmed and worshiped at organized centers. At Çatal Hüyük in Anatolia, shrines dating as far back as 8000 B.P. have been found in which clay and plaster reliefs of bulls' and rams' heads, some incorporating actual horns from the animals, occupy prominent positions.

The feelings of a farmer for his stock today rarely include religious awe, to be sure. But for all the complicating factors, from ego to economics, that are wrapped up in the modern farmer's concern for the well-being of his stock, there is a simple sincerity of feeling for one's animals that every farmer knows. E. B. White may have been more a writer than a farmer, and I don't know many farmers who talk this way, but I know many who would immediately recognize the sentiment he expressed in watching his pig grow ill:

> From the lustiness of a healthy pig a man derives a feeling of personal lustiness; the stuff that goes into the trough and is received with such enthusiasm is an earnest of some later feast of his own, and when this suddenly comes to an end and the food lies stale and untouched, souring in the sun, the pig's imbalance becomes the man's, vicariously, and life seems insecure, displaced, transitory.

White captured more than just the compassion in this ironically tinged passage; he captured that sense of greater awareness, that mastery of what is to the Peter Singers of the world a contradiction: If the pig recovers, it is going to die to feed us, but even so its illness and suffering is inseparable from our own. This is not just one introspective writer's philosophizing. When the pig finally dies, all of White's neighbors make it clear that they understand what he has gone through and feels: "The news of the death of my pig traveled fast and far, and I received many expressions of sympathy from friends and neighbors, for no one took the event lightly and the premature expiration of a pig is, I soon discovered, a departure which the community marks solemnly on its calendar, a sorrow in which it feels fully involved."

The ability to reconcile compassion or even reverence on

the one hand with the strength of mind to slaughter on the other comes directly and powerfully from an instinctive understanding that nature is something larger than us, or rather that in our practice of agriculture we are a part of nature. The hunter really *does* understand something that the average apartment-dweller does not. He has seen first-hand the machinery that supports life on this earth; he has become, if only for an afternoon, an active part of it rather than a passive recipient. The same is true—perhaps more true—for a farmer who has slaughtered one of his own animals. There is much anecdotal evidence that points to this deeper understanding that comes naturally to the farmer; there is even more serious psychological research that has been carried out. Harold Herzog, a psychologist at Western Carolina University, studied the attitudes of, as he put it, "a group of relatively naive students" who worked on a beef and hog farm run by Warren Wilson College, a small liberal-arts school in western North Carolina. The students worked fifteen hours per week in exchange for room and board; they cared for animals and helped in slaughtering. The results were surprising, precisely the opposite of what animal rightists claim would be our reaction if once we were all confronted with the gruesome realities of where our meat comes from:

> The students overwhelmingly, with only a couple of exceptions, felt that slaughtering had been a valuable and significant experience in their lives. . . . In general, the students felt that their farm crew experience had helped them clarify their attitudes about eating other creatures. They felt they were special. They knew things that others do not know—such as where meat comes from and what a warm carcass feels like. They had, at

least once in their lives, assumed the moral responsibility of carnivory.

Both the hunter and the farmer have learned something
about themselves; they have reforged a connection with
their cultural and historical roots—an idea that Leopold
embodies in his term "split-rail values" and Thoreau in his
reverence for the "primitive rank and savage" life—but
they have also learned something about the larger work-
ings of nature.

In raising animals, we are reenacting something not as
old, culturally speaking, as hunting, but in a way more
profound, for the rise of animal agriculture is an example
of evolution operating at its highest level—on *systems* of
species, one of which is us. It is much easier to suggest
doing away with animal agriculture if we subscribe to the
idea that it was all just another exploitive invention of man.
The realization that it is the product of something much
larger than us, a process that in fact substituted mutual
dependence for outright competition or predation, is a little
more humbling.

There is, admittedly, a danger in this line of argument:
If we aren't careful, we can end up claiming that *everything*
in human culture is nothing but the product of evolution,
so we daren't tamper with it. The Social Darwinists, who
assumed that all of human behavior was a product of an
individual's genetic makeup, and thus that all of society
was but a direct product of biological evolution, were the
most notorious proponents of this sort of reasoning. Thus
they argued that social progress was driven by a competition between the fit and the less fit, and, since this all is a
purely natural process, it would be subverted if social institutions interfered—by aiding the poor, for example.

Variations on this argument were used to justify the most reprehensible forms of racism in the late nineteenth century. Substitute "God" for "evolution," and you also have the justification invoked for every repression of man and exploitation of nature committed by man from feudal serfdom to Manifest Destiny. The mere fact that something is (for example, poverty) doesn't make it morally right; the mere fact that a practice invokes our cultural roots (for example, the genocide of the Indians) doesn't make it desirable.

By the same reasoning, one could argue (and animal rightists do) that just because domestic animals exist and just because our ancestors farmed, that is no justification for continuing to exploit them if doing so offends our moral sensibilities. But this objection blurs the important distinction between what is truly the product of evolution—that is, the natural selection of genetic traits that confer an adaptive advantage—and what is truly cultural. It is a distinction that we intuitively understand. Even though culture may be to some degree beyond our control (no one would claim that the Yanomamo Indians, for example, set out deliberately to invent a society guided by rules such as [1] wear loincloths, [2] manufacture clubs and bows and arrows, [3] kill someone if your mother dies), we generally recognize that human culture is ultimately subordinate to human will. If someone said we shouldn't get rid of nuclear weapons because they are the natural product of evolution, he would be dismissed as a crank. We know that we can by act of Congress eliminate slavery or by force of a treaty eliminate intermediate-range nuclear weapons.

The genetic character and behavior of domesticated animals, on the other hand, is arguably much more the product of evolution in its truest sense, something that is not subordinate to human consciousness. The fate of these species was dictated by nature more than by man's cultural institutions. By our actions we clearly acknowledge that nature—at least "wild" nature—is something different, something that lives by its own rules, something to which we cannot, and even should not, apply the standards of our own societies. Michael Fox may be guilty of slightly anthropomorphizing the realty when he says the deer "expects" to be killed by the wolf, but the larger point is undeniable. The wolf and the deer, by virtue of millions of years of evolution, are adapted in a way that their populations depend upon one another for survival; we disturb that relationship at the peril of causing ever greater disruptions to the ecosystem, which indeed are quite demonstrable in this particular case. The elimination of wolves has led to an explosion of deer populations in many parts of the United States; the deer strip bare hillsides and ranges that may never grow back, and many deer starve to death.

If we recognize what the scientific evidence is telling us about domestication, then man, in his practice of animal agriculture if nothing else, is as surely a product of nature. It is stretching the point to say that, because domestic sheep are a product of evolution, we daren't reduce our herds lest evolution's sublime ecological balance be destroyed. But that misses the point, anyway. Why do we really care whether wolves as a species, or prairie grass as a species, or lichen as a symbiotic pair of species survive? To the sentimental, who care about wolves but not lichen, and whales but not tree frogs, the answer is purely aesthetic. The more sophisticated defenders of endangered species and the "natural balance" spin utilitarian arguments to appeal to politicians and businessmen (useful drugs from tropical plants are a favorite). But what truly motivates them is a sense of reverence and awe at the sublime process of evolution that has created these species over thousands and millions of years. It is that sense of connection with the past and with a force much larger than ourselves that is at the core of the conservation ethic. It is that recognition that nature has its own laws that inspires us to be humble in imposing our own.

What farmers and hunters know by experience and instinct is something that others can appreciate through science and contemplation. Leopold, in a phrase that neatly expresses a wonder of nature while deflating the overromanticized pretenses of those who see nothing but ugliness and corruption in science and technology, observes that "wild things" had "little human value until mechanization assured us of a good breakfast, and until science disclosed the drama of where they came from and how they live." When we begin to appreciate the drama of where dogs, horses, and sheep came from, these "wild things" too attain a value that is all too often taken for granted. When we understand that farm animals are dependent upon us for their very survival by virtue of their genetic nature, by a genetic nature whose die was cast long before we began to practice conscious selective breeding, we develop a sense of obligation that is easy to dismiss if we ascribe their existence merely to man's conquest. Understanding that the domestic alliance is an evolutionary strategy of adaptive significance, that animals "chose" us because we were a better deal in an evolutionary sense than life in the wild, inspires a feeling of a bond between species that no amount of sentimental dripping or philosophizing about abstract "rights" can. Looking at the family dog lying on the living room floor, we can either see what most do—that is, almost nothing except a commonplace fixture in the twentieth-century American home—or we can see a link with the very dawn of civilization, when a series of evolutionary and climatic forces brought together two species in a way that transformed the face of the planet. The drama that the science of coevolution has disclosed may be able to inspire a sense of wonder and respect for animal life in a way that the airy, strident, and absolutist stance of animal rightists fails to.

The other humbling lesson that nature can teach us, if we let her, is the pretentiousness of our believing that we can even make clean moral choices about a system as complex as the natural world. Only someone who at least tacitly believes in the supreme power of humans to order the world to their making can believe in our ability to draw absolute moral lines, to say, If we do X, it will have consequence Y, and that is Right. Again, we all seem to *know* this when it comes to "wild" nature; we have learned the hard way that our attempts to manage wildlife by controlling predators or introducing new species more often than not fail—or worse. Yet the very people who so vociferously denounce man's attempt to extend his dominion over nature pretend that we can know with absolute moral certainty the outcome of our actions when it comes to our behavior toward domesticated animals. On the contrary, what little we do know shows that even when we are dealing in the realm of "tame" agriculture, we are tapping into a tangled web of interdependent, intertwining elements. Buy a cat (good), and you kill birds. Replace a cow pasture with an oat field (good), and you plow up a woodchuck's home and increase soil erosion and water pollution. Nature has its own tough laws that play havoc with any attempt to apply the moral code we have developed for our trade in human values. Freedom is an infinitely divisible commodity; habitat, especially on an earth filled already with five billion humans, is not. Nature is a world built upon compromise. The good we do is an approximation at best.

So where does this leave us? It does not mean we must abandon all hope of making any moral judgments about our dealings with animals. But it does suggest that there are fewer absolutes than we might like—certainly fewer than the animal rightists (who seem to conclude even the most scholarly of texts with a list of vegetarian recipes and boycott targets) would have us believe. One moral lesson that the natural historical view of domesticated animals can teach us is that just as sheep may "expect" to be killed in their symbiotic relationship with humans, they also, by virtue of their evolutionary "choice" thousands of years ago, "expect" to be better off than in the wild. That implies an obligation on our part to live up to our side of the bargain. However, one need not subscribe to the idea that animals have intrinsic rights to consistently oppose the infliction of needless cruelty, or to remember that we have a duty to animals who have cast their lot with ours and who by their evolutionary heritage are by nature dependent upon us.

A second point, however, is that the claims of what constitutes cruelty must be measured against a clear understanding of biology. Our urban isolation from the realities of nature have made us squeamish, or at least anthropomorphic; what may seem cruel may merely be the minimum necessary to prevent greater suffering (such as tail docking of sheep), or, as the story of the ewe who had to be confined to a crate illustrates, may be our anthropomorphic projections onto animals that are in fact genetically adapted to a life very different from what we simplistically imagine it to be. Biological studies can help us to clarify our thinking on these issues and separate our emotional prejudices from biological fact. The "choice" experiments mentioned in Chapter 7 are one empirical source of information about what animals themselves actually prefer.

Behavioral studies may likewise provide an objective basis for concluding that certain farming practices or animal experiments really do impose an unacceptable or avoidable degree of harm upon an animal's well-being, and for pointing the way to alternatives. Studies of pigs, for example, suggest that providing the animals access to ground that they can root in may substantially reduce stress and aggressive behavior in confinement.

More-careful behavioral and psychological study may also help provide a more scientific basis for drawing the distinctions between species that we all implicitly acknowledge exist. The wall that animal rightists attempt to erect between "sentient" and "nonsentient" beings is unidimensional as well as scientifically and philosophically problematic. It is difficult to understand why only animals that can feel pain should have an instrinsic right to live out their lives without interference from man. A more complete understanding of animals' innate capacities may, however, give us a more realistic sense of what an animal's "expectations" truly are. They surely differ from sheep to chimpanzee, for example. To take an exaggerated hypothetical example, but one that makes the point: If a livestock animal were capable of knowing throughout his life that he is to be killed and eaten, I would have no hesitation in saying that it would be morally wrong for us to continue the practice. Sheep do not have a capacity that approaches this. They can look at a dead companion apparently with total indifference. Chimpanzees, however, clearly have a sense of self-awareness and probably some sense of the future, a factor that ought to be included in our calculation of the moral cost of confining them to laboratory cages and using them as biomedical research subjects.

This is not a neat and clean answer. It does not have the virtue of simplicity that the absolutist position of the animals rightists offers. It can not be reduced to slogans or formulas. It is, like all moral dilemmas, a matter of balancing interests that conflict—often in agonizing ways. It requires knowledge, contemplation, and above all goodwill.

The case of animal experimentation is especially difficult. In the current climate of confrontation, the hope that knowledge and goodwill will someday reign seems forlorn at best. With both sides engaging in tactical maneuvering on legal and regulatory fronts, science and ethics have gone out the window; mature reflection is replaced by sloganeering. Animal activists won a series of legislative victories in the 1960s and 1970s requiring improvements, arguably much needed, in the standards for care of laboratory animals, federal inspections of research facilities, and the establishment of local committees to review research protocols that involve the use of animals. Now they have initiated a more concerted effort to attack the use of animals in research per se. The rhetoric of their campaign admits of no troubling complications. "No matter what the excuse, animal research is inexcusable," reads one advertisement from the National Anti-Vivisection Society. The text includes the blanket statement: "Animal research is obsolete. Alternative methods are better and cheaper."

Some of this is the result of either an ignorant or a calculated confusion of research with testing. Alternatives to the routine use of animals in tests, used to establish the safety of consumer products, are indeed being developed. The massive use of mice, rats, and rabbits, principally, in screening substances for carcinogenicity and toxicity may well be replaced by bacterial or cell culture assays. In all such tests, though, the quantity being measured is clear, the possible outcomes limited to well-defined possibilities.

Research is quite another matter. It is by definition an expedition into the unknown. In 1953, when Jonas Salk developed the first vaccine against polio, no cell culture or computer model could have predicted whether it was effective and safe—no one had ever made such a vaccine before.

Some of the antiresearch rhetoric goes much further, insisting against all fact that animal research is scientifically unnecessary or even misleading, since animals differ from humans and what is learned about the physiology of animals has no relevance to humans. Polio is again a good counter example: Monkeys are similar enough to humans that they can contract the polio virus. And a vaccine that worked in monkeys worked in people. Many surgical procedures, including open-heart surgery which has made it possible to correct congenital, and otherwise debilitating or fatal heart defects in infants, were developed and perfected in animals. No cell culture or computer model can tell a researcher what happens when he cuts into a beating heart.

The insistence that research is useless has become almost a religion for some activists; I have even seen a passionate, detailed, and utterly fantastic argument that the polio vaccine played no role in stemming the polio epidemic since the disease was naturally waning on its own.

Some protestors frankly admit that their object in seeking new regulations is to drive up the cost of animal research as high as possible; without such pressure, they say, researchers will have no incentive to seek alternatives. In response, researchers, arguing that free scientific inquiry can survive only if they are not continually in fear of being second-guessed by committees or legislatures, have dug in their heels, saying that the reforms they have accepted in terms of standards for humane care and facilities are as far as they can go. They are especially adamant that their judgment as to when an experiment requires the use of animal subjects not to be subject to external scrutiny. At the same time, the stridency of the opposition has made it all too easy for the research community to dismiss any criticism of animal experimentation as ignorant, if not downright misanthropic. Backed into such a defensive crouch, sophisticated contemplation of the moral costs of their work is not a top priority. Animals tend to become just pieces of lab equipment. In a climate where one is seen as either "for" or "against" animals, there is no room for saying that there is a moral cost when we kill an animal, but there are times when that moral cost is justified. The use of animals in the laboratory clearly offers a spectrum of benefits from the trivial to the profound (routine testing of the eye-irritating properties of cosmetics or the dissection of frogs in high school biology classes at one end of the spectrum, to developing techniques for the bone-marrow transplants that can save the lives of childhood lymphoma patients at the other); the ethical costs range across a spectrum, too, depending on the behavioral characteristics of the animal and the pain inflicted.* Finding that balance

---

* A 1988 survey by the Department of Agriculture, which enforces laboratory-animal regulations, found that 94 percent of lab animals are not exposed to painful procedures or are given pain-killing drugs.

will have to be the work of us all. Only through a better appreciation of man's and animals' place in nature will it ever be possible to face up to the fact, simple in itself but agonizing in its complications, that there is not necessarily a contradiction between using an animal for our justifiable ends while respecting it and fulfilling our responsibilities.

We as a society have lost the experiences in daily life that teach the moral lessons of such a natural ethic. An understanding of the scientific truths of how domesticated animals came to be can be the groundwork of a mature moral sense in a modern world that has otherwise lost the battle against falseness and unreality. It is only in that way that we may be able to bring a respect for the shared heritage of man and the other animals on this earth to bear upon the perpetual problem that nature has posed for us in managing the inevitably conflicting interests of earth's inhabitants.

REFERENCE LIST

This is not a complete bibliography, but only a list of the sources of quotations and data actually cited in each chapter. For works referred to in several places in the text, a reference is included only under the chapter where the first citation is found.

I. VISIONS OF NATURE

BADEN, JOHN A., AND DONALD LEAL, EDS. *The Yellowstone Primer: Land and Resource Management in the Greater Yellowstone Ecosystem*. San Francisco: Pacific Research Institute for Public Policy, 1990.

CHASE, ALTON. *Playing God in Yellowstone: The Destruction of America's First National Park*. Boston: Atlantic Monthly Press, 1986.

FRANCIS, DAVID R. "Fur Flies Over Wild Animal Trapping," *Christian Science Monitor,* April 4, 1988.

KLINGHAMMER, ERICH. "The Wolf: Fact and Fiction." In *Perceptions of Animals in American Culture,* ed. R. J. Hoage. Washington, D.C.: Smithsonian Institution Press, 1989.

PERLEZ, JANE. "A Puzzle for Zimbabwe: Too Many Elephants." *New York Times,* November 14, 1989.

———. "Where Elephants Roam, a Plea for Understanding." *New York Times,* August 9, 1989.

REDDING, RICHARD W. "A General Explanation of Subsistence Change: From Hunting and Gathering to Food Production." *Journal of Anthropological Archaeology,* Vol. 7 (1988), pp. 56–97.

RITVO, HARRIET. *The Animal Estate.* Cambridge, Mass: Harvard University Press, 1987.

SATCHELL, MICHAEL. "The American Hunter Under Fire." *U.S. News & World Report,* February 5, 1990, pp. 30–36.

———. "Yellowstone Lives." *U.S. News & World Report,* May 15, 1989, pp. 24–26.

SHABECOFF, PHILIP. "Seeing Disaster, Groups Ask Ban on Ivory Imports." *New York Times,* June 2, 1989.

SINGER, PETER. *Animal Liberation.* New York: New York Review, 1975.

WEBB, SARA. "Swedes Pay Price of Animal Rights." *Financial Times,* September 7, 1988.

II. CIVILIZATION'S PROGRESS;
OR, WHO INVENTED THE DOG?

DANIEL, GLYN. *The Origins and Growth of Archaeology.* New York: Thomas Y. Crowell, 1967.

DIAMOND, JARED. "The Worst Mistake in the History of the Human Race." *Discover,* May 1987, pp. 64–66.

GLASS, BENTLEY, ET AL., EDS. *Forerunners of Darwin.* Baltimore: Johns Hopkins Press, 1959.

HOLMES, JOHN. *The Farmer's Dog.* London: Popular Dogs, 1982.

LEGGE, ANTHONY J., AND PETER A. ROWLEY-CONWY. "Gazelle Killing in Stone Age Syria." *Scientific American,* August 1987, pp. 88–95.

LEWIN, ROGER. "A Revolution of Ideas in Agricultural Origins." *Science,* Vol. 240 (1988), pp. 984–86.

McELVAINE, ROBERT. *The Great Depression.* New York: Times Books, 1984.

McFARLAND, DAVID, ED. *The Oxford Companion to Animal Behavior.* Oxford, U.K.: Oxford University Press, 1987.

RINDOS, DAVID. *The Origins of Agriculture.* Orlando, Fla.: Academic Press, 1984.

ROUSE, IRVING. *Introduction to Prehistory.* New York: McGraw-Hill, 1972.

ROWAN, ANDREW N. "The Power of the Animal Symbol and its Implications." In *Animals and People Sharing the World,* ed. Andrew N. Rowan. Hanover, N.H.: University Press of New England, 1988.

WELLS, KEN. "In the Funky Streets of Berkeley, Calif., It's Forever the '60s." *Wall Street Journal,* January 11, 1989.

ZVELEBIL, MAREK. "Postglacial Foraging in the Forests of Europe." *Scientific American,* May 1986, pp. 104–15.

III. THE VIRTUES OF DEFENSELESSNESS

CLUTTON-BROCK, JULIET. *Domesticated Animals from Early Times.* Austin: University of Texas Press, 1981.

DIAMOND, JARED. "Strange Traveling Companions." *Natural History,* December 1988, pp. 22–27.

HART, BENJAMIN L. *The Behavior of Domestic Animals.* New York: W.H. Freeman, 1985.

ISACK, H.A., AND H. U. REYER. "Honeyguides and Honey

Gatherers: Interspecific Communication in a Symbiotic Relationship." *Science,* Vol. 243 (1989), pp. 1343–46.

KREBS, J. R., AND N. B. DAVIES, EDS. *Behavioural Ecology: An Evolutionary Approach.* Sunderland, Mass.: Sinauer Associates, 1984.

ROBINSON, MICHAEL. *Smithsonian,* February 1989, pp. 38–40.

STOKES, DONALD, AND LILLIAN STOKES. *Animal Tracking and Behavior.* Boston: Little, Brown, 1986.

TINBERGEN, NIKOLAAS. *The Study of Instinct.* Oxford, U.K.: Oxford University Press, 1951.

IV. THE SPECIES THAT CAME IN FROM THE COLD

CARSON, HAMPTON L. "The Process Whereby Species Originate." *Bioscience,* Vol. 37 (1987), pp. 715–20.

FOGG, JOHN M., JR. *Weeds of Lawn and Garden.* Philadelphia: University of Pennsylvania Press, 1956.

GEIST, VALERIUS. *Mountain Sheep: A Study in Behavior and Evolution.* Chicago: University of Chicago Press, 1971.

MASON, IAN L., ED. *The Evolution of Domesticated Animals.* London: Longman, 1984.

PIGGOTT, STUART. *Ancient Europe.* Chicago: Aldine, 1965.

SCHNEIDER, STEPHEN H., AND RANDI LONDER. *The Coevolution of Climate and Life.* San Francisco: Sierra Club Books, 1984.

SMITH, BRUCE D. "Origins of Agriculture in Eastern North America." *Science,* Vol. 259 (1989), pp. 1566–70.

TIMM, ROBERT M., AND JAMES S. ASHE. "The Mystery of the Gracious Hosts." *Natural History,* September 1988, pp. 6–10.

V. YOUTHFUL DESIGNS

COPPINGER, LORNA, AND RAYMOND P. COPPINGER. "Livestock-Guarding Dogs that Wear Sheep's Clothing." *Smithsonian,* April 1982, pp. 64–73.

———. "So Firm a Friendship." *Natural History,* March 1980, pp. 12–26.

COPPINGER, RAYMOND P., AND CHARLES K. SMITH. "The Domestication of Evolution." *Environmental Conservation,* Vol. 10 (1983), pp. 283–92.

COPPINGER, RAYMOND P., CHARLES K. SMITH, AND L. MILLER. "Observations on Why Mongrels Make Effective Livestock-Protecting Dogs." *Journal of Range Management,* Vol. 38 (1985), pp. 560–61.

COPPINGER, RAYMOND P., AND LORNA COPPINGER. "Livestock-Guarding Dogs." *Country Journal,* April 1980, pp. 68–77.

GOULD, STEPHEN J. "A Biological Homage to Mickey Mouse." In *The Panda's Thumb.* New York: W.W. Norton, 1980.

LAWRENCE, ELIZABETH A. "Neoteny in American Perceptions of Animals." In *Perceptions of Animals in American Culture,* ed. R. J. Hoage. Washington, D.C.: Smithsonian Institution Press, 1989.

SIMMONS, PAULA. *Raising Sheep the Modern Way.* Charlotte, Vt.: Garden Way Publishing, 1976.

VI. NO TURNING BACK

COLE, JOHN. *Archaeology by Experiment.* New York: Charles Scribner's Sons, 1973.

HOWELL, JOHN M. "Early Farming in Northwestern Europe." *Scientific American,* November 1987, pp. 118–26.

YELLEN, JOHN E. "The Transformation of the Kalahari Kung." *Scientific American,* April 1990, pp. 96–105.

VII. HUMAN PIETIES

ATKINSON, B. A., AND V. LORIAN. In *American Society of Microbiology Meeting,* March 7, 1984.

CAUDILL, HARRY M. *Night Comes to the Cumberlands.* Boston: Little, Brown, 1962..

CHAGNON, NAPOLEON. "Life Histories, Blood Revenge, and Warfare in a Tribal Population." *Science,* Vol. 239 (1988), pp. 985–92.

CHURCHER, PETER B. AND JOHN H. LAWTON. "Beware of Well-fed Felines." *Natural History,* July 1989, pp. 40–47.

DOBSON, JESSIE. In *Diseases in Ancient Man,* ed. Gerald D. Hart. Toronto: Clarke Irwin, 1983.

EHRLICH, PAUL R., AND ANNE H. EHRLICH. *The Population Explosion.* New York: Simon and Schuster, 1990.

FOX, MICHAEL W. *Returning to Eden.* New York: Viking, 1980.

————. *Farm Animals: Husbandry, Behavior, and Veterinary Practice.* Baltimore: University Park Press, 1984.

————. *Inhumane Society: The American Ways of Exploiting Animals.* New York: St. Martin's Press, 1990.

MAY, ROBERT M. "Control of Feline Delinquency." *Nature,* Vol. 332 (1988), pp. 392–93.

MURO, MARK. "When Animal Rights Go Wrong." *Boston Globe,* October 30, 1988.

NEIHARDT, JOHN G. *Black Elk Speaks.* New York: William Morrow, 1932.

SAPOLSKY, ROBERT S. "Stress in the Wild." *Scientific American,* January 1990, pp. 116–23.

SCOTT, GEORGE E. *The Sheepman's Production Handbook.*

Denver: Sheep Industry Development Program, 1970.

STEADMAN, DAVID W., AND STORRS L. OLSON. "Bird Remains from an Archeological Site on Henderson Island, South Pacific: Man-Caused Extinctions on an 'Uninhabited' Island." *Proceedings of the National Academy of Science,* Vol. 82 (1985), p. 6191.

TWAIN, MARK. *Letters from the Earth.* New York: Harper & Row, 1962.

WADE, DALE A., AND JAMES E. BROWNS. *Procedures for Evaluating Predation on Livestock and Wildlife.* Bulletin B-1429, Texas Agricultural Extension Service.

Wild Animals of North America. Washington, D.C.: National Geographical Society, 1979.

VIII. A NATURAL ETHIC

CAMPBELL, JOSEPH. *The Power of Myth.* New York: Doubleday, 1988.

HERZOG, HAROLD A., JR., AND GORDON M. BURGHARDT. "Attitudes Toward Animals: Origins and Diversity." In *Animals and People Sharing the World,* ed. Andrew N. Rowan. Hanover, N.H.: University Press of New England, 1988.

LEOPOLD, ALDO. *A Sand County Almanac.* New York: Oxford University Press, 1949.

LOBBAN, RICHARD A., JR. "Cattle and the Rise of the Egyptian State." *Anthrozoos,* Vol. 2 (1989), pp. 104–201.

MELLAART, JAMES. *Earliest Civilizations of the Near East.* London: Thames and Hudson, 1965.

RITVO, HARRIET. "The Emergence of Modern Pet-Keeping." In *Animals and People Sharing the World,* ed. Andrew N. Rowan. Hanover, N.H.: University Press of New England, 1988.

ROBERTSON, R. B. *Of Sheep and Men.* New York: Alfred A. Knopf, 1957.

SPERLING, SUSAN. *Animal Liberators.* Berkeley: University of California Press, 1988.

THOREAU, HENRY DAVID. *Walden.* 1854.

U.S. DEPARTMENT OF AGRICULTURE. *Animal Welfare Enforcement Report—Fiscal Year 1988.* Washington, D.C.: USDA, 1989.

VAN DER POST, LAURENS. *A Story like the Wind.* New York: William Morrow, 1972.

WHITE, E. B. "Death of a Pig." In *Essays of E. B. White.* New York: Harper & Row, 1977.

INDEX